INTEGRATED MULTIREGION MODELS
FOR POLICY ANALYSIS:
AN AUSTRALIAN PERSPECTIVE

Studies in Regional Science and Urban Economics

Series Editors

ÅKE E. ANDERSSON
WALTER ISARD
PETER NIJKAMP

Volume 19

NORTH-HOLLAND – AMSTERDAM • NEW YORK • OXFORD • TOKYO

Integrated Multiregion Models for Policy Analysis: An Australian Perspective

CHRISTINE SMITH
Griffith University
Nathan, Queensland
Australia

1989

NORTH-HOLLAND –AMSTERDAM ● NEW YORK ● OXFORD ● TOKYO

© ELSEVIER SCIENCE PUBLISHERS B.V., 1989

All rights reserved. No part of this publication may be reproduced, stored in a retrieval system, or transmitted in any form or by any means, electronic, mechanical, photocopying, recording or otherwise, without the prior written permission of the publisher, Elsevier Science Publishers B.V. (Physical Sciences and Engineering Division), P.O. Box 1991, 1000 BZ Amsterdam, The Netherlands.

Special regulations for readers in the USA - This publication has been registered with the Copyright Clearance Center Inc. (CCC), Salem, Massachusetts. Information can be obtained from the CCC about conditions under which photocopies of parts of this publication may be made in the USA. All other copyright questions, including photocopying outside of the USA, should be referred to the copyright owner, Elsevier Science Publishers B.V., unless otherwise specified.

No responsibility is assumed by the publisher for any injury and/or damage to persons or property as a matter of products liability, negligence or otherwise, or from any use or operation of any methods, products, instructions or ideas contained in the material herein.

ISBN: 0 444 87155 1

Published by:

ELSEVIER SCIENCE PUBLISHERS B.V.
P.O. Box 1991
1000 BZ Amsterdam
The Netherlands

Sole distributors for the U.S.A. and Canada:

ELSEVIER SCIENCE PUBLISHING COMPANY, INC.
655 Avenue of the Americas
New York, N.Y. 10010
U.S.A

Library of Congress Cataloging-in-Publication Data

```
Smith, Christine, 1956-
    Integrated multiregion models for policy analysis : an Australian
 perspective / Christine Smith.
       p.    cm. -- (Studies in regional science and urban economics ;
  v. 19)
    Bibliography: p.
    ISBN 0-444-87155-1
    1. Regional planning--Mathematical models.  2. Regional planning-
 -Australia--Mathematical models.  3. Regional economics-
 -Mathematical models.  4. Input-output analysis.  I. Title.
  II. Series.
  HT391.S565   1989
  338.994--dc19                                              88-39112
                                                                 CIP
```

PRINTED IN THE NETHERLANDS

ACKNOWLEDGEMENTS

This book is an outgrowth of my doctoral dissertation completed in the field of regional scienced at Cornell University. In the process of working on this book I have incurred a number of debts -- both intellectual and material.

Over a decade ago, Rod Jensen awakened my interest in the field of regional science, and suggested subsequently that I explore the possibility of further study in the United States. He continues to challenge me to prove the empirical validity of my research endeavours, and for this I am extremely grateful.

A more significant debt is owed however to Walter Isard. Walter has fulfilled the multiple roles of professor, thesis advisor and inspirational mentor over many years now. His boundless energy, enthusiasm and creative insights continue to make dramatic impressions on my own writing and thinking. It is to him that this book is dedicated.

I wish to acknowledge also the guidance and encouragement provided to me by the other members of my PhD committee at Cornell -- Professors Stan Czamanski and William Lucas.

I have been fortunate to receive considerable financial assistance during the period of this research. For their generous support, I am indebted to Cornell University, the Queensland government, the University of Queensland, the World Research Centre Incorporated, the University of Wollongong, Griffith University, and the Federal government sponsored Australian Research Grants Scheme.

Finally, I would like to thank my family for their constant support and encouragement, Helena Wood for her cheerful perserverance in typing the early versions of the manuscript despite being a continent or two away in terms of geographic location, Olwen Schubert for her painstaking efforts in producing a camera-ready version of the manuscript, and the reviewers of the original manuscript (Professors Peter Nijkamp and Piet Rielveld) whose insightful comments have helped to improve the quality of the final manuscript considerably.

Christine Smith
August 1988

TABLE OF CONTENTS

CHAPTER 1. INTRODUCTION   1

   1.1 Background and Objectives of the Study   1
   1.2 Outline of the Study   2

**CHAPTER 2. MULTIREGIONAL ECONOMIC BASE AND
INPUT-OUTPUT MODELS**   7

   2.1 Introduction   7
   2.2 Economic Base Models   7
   2.3 Regional Input-Output Models   9
   2.4 Balanced Regional Input-Output Models   13
   2.5 Interregional Input-Output Models   18
      2.5.1 Isard's Ideal Type   18
      2.5.2 The 'Fixed Shares of Total Purchases from
each Region' Type   23
      2.5.3 The 'Fixed Shares of Total Sales to each
Region' Type   27
      2.5.4 Leontief-Strout and Other Gravity-Type Models   29
      2.5.5 Polenske's MRIO Project   36
   2.6 Concluding Remarks   37

**CHAPTER 3. A COMPARATIVE COST AND INDUSTRIAL
COMPLEX MODULE AS A COMPONENT OF AN
INTEGRATED MULTIREGIONAL MODEL**   45

   3.1 Introduction   45
   3.2 Outline of Comparative Cost and Industrial Complex
Analysis   45
      3.2.1 Comparative Cost Approach   45
         a. Single Region Applications   46
         b. Multiregion Applications   54
         c. Evaluation of the Approach   60
      3.2.2 Industrial Complex Approach   62
         a. Single Region Applications   63
         b. Multiregion Applications   74
         c. Evaluation of the Approach   76
   3.3 Integration of Comparative Cost-Industrial Complex
Analysis and Multiregional Input-Output Models   78
   3.4 Concluding Remarks   86

APPENDIX 3A: EXTENSION OF COMPARATIVE COST AND

INDUSTRIAL COMPLEX ANALYSIS TO
INTERNATIONAL APPLICATIONS   96

CHAPTER 4. A LINEAR PROGRAMMING MODULE AS A COMPONENT OF AN INTEGRATED MULTIREGIONAL MODEL   97

4.1 Introduction   97
4.2 Outline of Multiregional Linear Programming Models   98
    4.2.1 Structure of a General Interregional Programming Model   99
    4.2.2 Individual Sector Interregional Programming Models   103
4.3 Integration of Multiregional Linear Programming and Input-Output Models 107
    4.3.1 Input-Output Relations as Constraints in Programming Models 107
        a. The REGAL (REGional ALlocation of all productive sectors) Model 110
        b. The MORSE (MOdel for the analysis of Regional development, Scarce resources and Employment) Model 115
    4.3.2 Linkage of Multiregional Linear Programming and Input-Output Modules 121
4.4 Integration of Comparative Cost-Industrial Complex, Programming and Input-Output Models 127
    4.4.1 Results of Comparative Cost-Industrial Complex Analysis as Constraints in Programming Models 127
    4.4.2 Linkage of Comparative Cost-Industrial Complex, Input-Output and Programming Models 132
4.5 Concluding Remarks   137

APPENDIX 4A: MULTIREGIONAL LINEAR PROGRAMMING MODELS OF AUSTRALIAN AGRICULTURE 144

CHAPTER 5. A DEMOGRAPHIC MODULE AS A COMPONENT OF AN INTEGRATED MULTIREGIONAL MODEL 149

5.1 Introduction   149
5.2 Outline of Multiregional Demographic Models   149
    5.2.1 Simple Components of Change Approach   149
    5.2.2 Cohort-Survival Approach   152
    5.2.3 Microsimulation Approach   157
5.3 Alternative Approaches to Modelling Migration   160

	5.3.1	Gravity Model Approach	160
		a. Unconstrained Model for Aggregate Migration	160
		b. Unconstrained Model for Migration by Demographic Class	165
		c. Doubly Constrained Models	167
		d. Other Constrained Gravity-Type models	171
	5.3.2	Econometric Approach	174
		a. Inmigration Equations	176
		b. Outmigration Equations	179
		c. Interregional Migration Flows	179
5.4	Integration of Multiregional Input-Output and Demographic Models		180
5.5	Integration of Multiregional Input-Output, Comparative Cost-Industrial Complex, Programming and Demographic Modules		186
5.6	Concluding Remarks		192

CHAPTER 6. A MULTIREGIONAL ECONOMETRIC MODULE AS A COMPONENT OF AN INTEGRATED MULTIREGIONAL MODEL   207

6.1 Introduction   207
6.2 Outline of the Multiregional Econometric Module   207
    6.2.1 Household Income and Consumption Expenditure Submodel   208
       a. Household Income Equations   208
       b. Household Consumption Expenditure Equations   213
    6.2.2 Labour Market Submodel   215
       a. Employment Demand Equations   215
       b. Labour Supply Equations   217
       c. Wage Rate Equations   220
    6.2.3 Private Investment Expenditure Submodel   222
       a. Agricultural, Manufacturing and Mining Sectors   222
       b. Residential Construction Sector   224
       c. Other Sectors   225
    6.2.4 Government Expenditure Submodel   225
       a. Federal Government Expenditure Equations   227
       b. State and Local Government Current Expenditure Equations   227
       c. State and Local Government Capital Expenditure Equations   230
6.3 Integration of Multiregional Input- Output and Econometric Modules   231
    6.3.1 Linkage with Household Income and Consumption Expenditure Submodel   232
    6.3.2 Linkage with Government Expenditure Submodel   234
    6.3.3 Linkage with Labour Market Submodel   235

6.3.4 Linkage with Private Investment Expenditure Submodel	236
6.4 Integration of Multiregional Input-Output, Demographic and Econometric Modules	238
6.5 Integration of Multiregional Input-Output, Comparative Cost-Industrial Complex, Programming, Demographic and Econometric Modules	240
6.6 Concluding Remarks	245

CHAPTER 7. A FACTOR DEMAND MODULE AS A COMPONENT OF AN INTEGRATED MULTIREGIONAL MODEL — 253

7.1 Introduction	253
7.2 Outline of Factor Demand Models	253
7.2.1 Translog Cost Function	254
7.2.2 Diewert (Generalized Leontief) Cost Function	257
7.2.3 Further Refinements	258
a. Recognition of Scale Economies	258
b. Recognition of Partial Adjustment Process	259
7.3 Relationship Between Factor Demand and Input-Output Models	260
7.3.1 Input-Output Model with Prices Explicit	260
a. Regional Input-Output Relations	261
b. Multiregional Input-Output Relations	264
7.3.2 Translog Cost Shares as Input-Output Coefficients	265
7.3.3 Diewert Derived Factor Demands as Input-Output Coefficients	266
7.4 Linkage of Factor Demand and Input-Output Models	268
7.4.1 Case of Aggregate MELK Model	268
a. Use of Translog Cost Function	268
b. Use of Diewert Cost Function	269
7.4.2 Case of Disaggregated $M(m_1,m_2,....)$, $E(e_1,e_2,....)$, L, K Model	270
a. Use of Across-the-Board Coefficient Adjustment	271
b. Use of Restricted Coefficient Adjustment	274
c. Use of Two-Level Optimization	276
7.5 Integration of Multiregional Econometric, Factor Demand and Input-Output Modules	278
7.6 Integration of Multiregional Demographic, Factor Demand and Input-Output Modules	281
7.7 Integration of Multiregional Input-Output, Comparative Cost-Industrial Complex, Programming, Demographic, Econometric and Factor Demand Modules	284
7.8 Concluding Remarks	294

CHAPTER 8. POTENTIAL EXTENSIONS TO INTEGRATED MODEL — 301

- 8.1 Introduction — 301
- 8.2 Inclusion of a National Econometric Module (NATLEC) — 301
- 8.3 Inclusion of a Transportation Module (TRANS) — 303
- 8.4 Evaluation of Alternative Solution Algorithms — 304
- 8.5 Concluding Remarks — 307

CHAPTER 9. POLICY ANALYSIS USING AN INTEGRATED MULTIREGION MODEL — 309

- 9.1 Introduction — 309
- 9.2 Illustrative Use of Policy Projection Framework — 309
  - 9.2.1 The Setting of Key Basic Forces: Historical Perspectives and Less Sophisticated Approaches — 310
    - a. Single Decision-making (Governmental) Unit — 310
    - b. Many Interest Groups Influencing a Single Decision-making (Governmental) Unit — 316
  - 9.2.2 A More Comprehensive and Systematic Approach Toward Policy Projection — 318
    - a. Basic elements of a policy choice using the Saaty Approach: Single Decision-making (Governmental Unit) — 319
    - b. Extension to Many Decision-making (Governmental) Units: Modifications to the Basic Elements of a Policy Choice — 328
  - 9.2.3 Difficulties and Qualifications in Policy Projection Framework — 330
- 9.3 Relationship of INPOL Module to other Modules Comprising the Integrated Model — 331
- 9.4 Concluding Remarks — 332

CHAPTER 10. CONCLUSIONS — 337

REFERENCES — 343

LIST OF TABLES

TABLE #

2.1	A Simplified Regional Transactions Table	10
3.1	Locationally Significant Costs per Unit Output of Industry i at Selected Locations	50
3.2	A Hypothetical Interactivity Matrix for Selected Input/Output Combinations	67
3.3	Total Requirements and Yields of Selected Industrial Complexes	71
3.4	Modified Interregional Input-Output Transactions Table	82
9.1	Key Policy Issue Pressure Indicators	311

LIST OF ILLUSTRATIONS

FIGURE #

1.1 Components of an Integrated Multiregional Model for Australia — 3

2.1 A Schematic Representation of the Balanced Regional Input-Output Model — 14

2.2 A Schematic Representation of Flows of Commodity i in the Leontief-Strout Gravity Model — 30

3.1 A Schematic Representation of the Comparative Cost Approach in Single Region Applications — 47

3.2 A Schematic Representation of the Comparative Cost Approach in Multiregion Applications — 55

3.3 A Schematic Representation of the Industrial Complex Approach in Single Region Applications — 64

3.4 Linkage of Comparative Cost-Industrial Complex and Interregional Input-Output Model — 79

4.1 A Schematic Representation of Paths Between Energy Resource Supplies and Functional Energy Service Demand Satisfaction in the Brookhaven Programming Model — 106

4.2 Structure of the MORSE Model — 116

4.3 Linkage of Multiregional Linear Programming and Input-Output Submodels — 124

4.4 Linkage of Comparative Cost-Industrial Complex and Multiregional Programming Analysis — 129

4.5 Linkage of Comparative Cost-Industrial Complex, Input-Output and Programming Submodels — 133

5.1 Linkage of Input-Output and Demographic Modules — 182

5.2 Linkage of Input-Ouptut, Comparative Cost-Industrial Complex, Programming and Demographic Modules — 187

6.1 The Household Income and Consumption Expenditure Submodule — 209

6.2 The Labour Market Submodule — 216

6.3 The Private Investment Expenditure Submodule — 223

6.4 The Government Expenditure Submodule — 226

6.5 Linkage of Input-Output, Comparative Cost-Industrial Complex, Programming, Demographic and Econometric Modules — 241

9.1 Steps Involved in Policy Projection Framework — 320

CHAPTER 1

INTRODUCTION

1.1 Background and Objectives of the Study

Over the past 25 years, a feature common to the regional science literature in many different countries has been the development of operational methods of regional analysis (eg. economic base, input-output, comparative cost-industrial complex, linear programming, demographic, econometric and transportation models). Three recent trends in this literature which provide the focus for this book are:

(1) an emphasis on multiregion rather than single-region model building, since this:

-explicitly recognizes the importance of interregional linkages,

-facilitates consideration of the geographic equity aspects of federal policies, and

-permits identification of the interregional spillover and/or feedback effects of policies adopted by a single state or local government unit.[1]

(2) an emphasis on integrated model building efforts as a result of a recognition that each type of multiregional model has a different set of strengths and weaknesses, and hence that benefits can be gained from a complementary rather than competitive use of such alternative types of models.[2]

(3) an emphasis on the need to design these models in such a way as to facilitate consistent analysis of policy choices and development options by local, state and federal governments and the private sector.[3]

The most ambitious integrated multiregion modelling effort is that proposed by Isard and Anselin (1982). However, some components of that model are not yet operational and will need to be revised given the paucity of the data available to support their implementation. Further, each of the modules comprising this integrated model has been developed separately by a different group of researchers -- and for the most part without a view to linkage with other modules. Hence, inconsistencies arise which will need to be ironed out before the entire model could be used. Finally, although Isard and Anselin (1982) identify the set of outputs from each model which will be fed to other modules within the integrated model, a sequence of steps was not specified by which to effect this linkage.

The objectives of this book are twofold:

(1) to specify an integrated multiregional model such that each component module is capable of implementation given the nature of data available, and

(2) to develop sequences of steps that are both reasonable and feasible for achieving the necessary linkages between these modules.

In order to ensure that these objectives could be realized, it was decided to focus on the Australian multiregional system. This was so because this system is the one which the author is most familiar with, and hence the one for which the full set of available data could most easily be assembled.

1.2 Outline of the Study

As depicted in Figure 1.1, the integrated model being proposed for implementation in the Australian context has seven major components: (1) an interregional input-output module, (2) a comparative cost-industrial complex module, (3) an interregional programming module, (4) a demographic module, (5) a regional econometric module, (6) a factor demand (substitution) module, and (7) a conflict management - multipolicy formation module.

INTRODUCTION

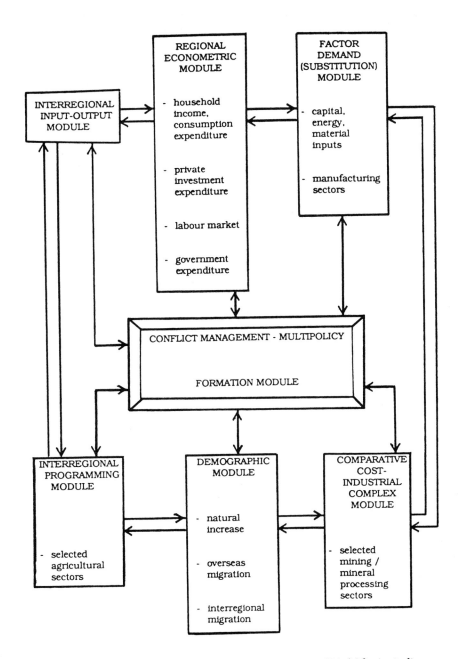

Figure 1.1 Components of an Integrated Multiregional Model for Australia

Interregional Input-Output Module

The core component of the integrated model is the one which determines (projects) the level of economic activity in each region. As discussed in Chapter 2, this component could comprise a multiregional economic base model, a set of regional input-output models, a balanced regional input-output model, or an interregional input-output model. However, the projections (output, employment, etc.) derived from an interregional input-output model can reasonably be expected to be the most accurate.

Comparative Cost-Industrial Complex Module

For those industries (or parts of industries) whose regional distributions are heavily influenced by transport costs, energy costs, and other Weberian-type locational factors the projections (output, employment, etc.) derived from the interregional input-output module can often be improved upon. In particular, the locational patterns of these industries -- primarily mining and/or mineral processing industries in the Australian context -- will be projected via the use of a comparative cost-industrial complex module of the form discussed in Chapter 3.

Interrregional Programming Module

Another set of industries for which input-output projections can be improved upon are those characterized by input and/or output substitution. In the Australian context, some agricultural sectors fall in this category since farmers in these sectors are able to adjust their output mixes (and hence input requirements) as a function of their expectations regarding relative output prices. An interregional programming module, of the general form discussed in Chapter 4, is included in order to handle such sectors adequately.

Demographic Module

Since regional demographic and economic projections can both be improved when made simultaneously, a demographic module is included as a basic component of the integrated model. As discussed in Chapter 5, this module contains: (1) a natural increase component, (2) an overseas migration component, and (3) an interregional migration component. Since the third component is of most interest to regional scientists, it is the one developed in the most detail in this chapter.

Regional Econometric Module

As discussed in Chapter 6, the regional econometric module proposed can be divided into four submodules:

(1) a household income and consumption expenditure submodule,
(2) a labour market submodule,
(3) a private investment expenditure module, and
(4) a government expenditure submodule.

Its role in the integrated model is to provide employment demand, labour supply and wage rate projections for each region, and to give an indication of revisions required in the consumption, investment and government expenditure components of the final demand vector employed when operating the interregional input-output module.

Factor Demand Module

When using an interregional input-output module for projection purposes, it is important to allow technical and (ideally) trade coefficients to change over time in response to technological innovations and changes in factor input prices. As a result, a factor demand (substitution) module which allows for substitution between capital, labour, energy and material inputs is included in the model. The form of this module is discussed in Chapter 7.

Other Non-Policy Modules and Solution Algorithms

In Chapter 8, extension of the integrated model depicted in Figure 1.1 to include other non-policy modules is considered. In particular, a national econometric module and a transportation module are discussed. However, the demonstration of the operationality of such modules and/or the development of their linkages to other components of the integrated model awaits further research. Some comments are also made in this chapter on the relative merits of different solution algorithms for operating the proposed model.

Conflict Management-Multipolicy Formation Module

Finally, in Chapter 9, it is recognized that regional changes are affected in a major way by the type of compromise policy mixes (sets of policies) adopted by political leaders taking into account their own conflicting objectives and those of the diverse interest groups comprising their constituencies. The integrated model is then extended to incorporate an additional module capable of forecasting the compromise policies most likely to be adopted.

NOTES

[1] For general reviews of multiregional model building efforts, see Adams and Glickman (1980), Bolton (1980a, 1980b, 1982a, 1982b), Courbis (1982b), Glickman (1982), Issaev, et al. (1982), Nijkamp and Rietveld (1980, 1982a, 1982b), Nijkamp, et al. (1982), Rietveld (1981, 1982a, 1982b) and Snickars (1982).

[2] For further discussion of this type of philosophy, see Isard, et al. (1960), Isard, et al. (1981), Isard and Anselin (1982), Isard and Smith (1983, 1984a, 1986), Lakshmanan (1982), Smith (1983), Treyz (1980) and Treyz and Stevens (1980).

[3] For further elaboration, see Bolton (1980a, 1982a, 1982b), Glickman (1982), Isard, et al. (1981), Isard and Smith (1984, 1984a, and 1984b), Nijkamp and Rietveld (1982b) and Smith (1983, pp. 478-571).

CHAPTER 2
MULTIREGIONAL ECONOMIC BASE AND INPUT-OUTPUT MODELS

2.1 Introduction

This chapter reviews those models which could be employed as the core module within an integrated multiregion model -- namely those models which are capable of determining the level of economic activity by sector in each region. More specifically, four types of models are considered:

(a) a multiregional economic base model in Section 2.2;

(b) a set of regional input-output models in Section 2.3;

(c) a balanced regional input-output model in Section 2.4; and

(d) an interregional input-output model in Section 2.5.

Since, generally speaking, an interregional input-output model can be expected to yield the most reasonable output projections a number of variants of this module are considered -- each employing a different set of assumptions when constructing interregional trade coefficients. Finally, a few concluding remarks are made in Section 2.6.

2.2 Economic Base Models

The economic base model is the oldest and simplest model employed by regional economists.[1] It is based on the division of each region J's economic activity into two mutually exclusive sectors:

(a) a basic (primary or export) sector comprising activity ^JX_B dependent on demand external to J; and

(b) a service (non-basic or residentiary) sector comprising activity $^JX_{NB}$ dependent on demand internal to J.

The level of activity in the service sector is then expressed as a function of that in the basic sector:

$$^JX_B = {}^Jf({}^JX_{NB}) \qquad (2.2.1)$$

On the assumption that the relationship in (2.2.1) is stable over time, it is possible to calculate a simple multiplier $^J k$, where:[2]

$$^J k = (^J X_B + {}^J X_{NB})/{}^J X_B \qquad (2.2.2)$$

This multiplier can then be used to forecast future levels of economic activity in region J, given projections of $^J X_B$.

Despite its widespread use, this model is subject to some major technical and conceptual difficulties, especially when applied simultaneously to a set of regions. Among these are:

(a) the choice of a method for allocation of activities between the basic and non-basic sectors, since each method currently in use has a different set of drawbacks;[3]

(b) the choice of the level of sectoral disaggregation to be employed when allocating sectors into basic/service components, since the results are highly sensitive to this choice;[4]

(c) the choice of a method for estimation of the regional multiplier $^J k$, since the interpretation of the results depends on the number of observations employed in its calibration;[5]

(d) a failure to differentiate between two distinct types of multiplier relationships, namely those resulting from interindustry interactions and household/industry interactions;[6]

(e) a failure to recognize the interdependence between different basic sectors, and inability to differentiate results according to which basic sector receives the initial injection;[7]

(f) an exclusive preoccupation with demand-side considerations;[8]

(g) a failure to recognize imports as a counterpart of exports;[9] and

(h) a failure to incorporate interregional relationships, and hence an inability to incorporate interregional 'feedback' effects in impact projections.

Despite modifications which may be made to the basic model presented above in an attempt to avoid one or more of these problems, too many remain for it to be considered for adoption should available data and research resources permit implementation of one or more of the approaches to be discussed below.

2.3 Regional Input-Output Models

The input-output model, as its name implies, is concerned with the inputs and outputs of each sector in an economy and their interrelationships. The structure of these interrelationships can be formally identified and recorded in tabular format. To illustrate, consider a regional economy J which is divided into n production sectors. Let $^J X_i$ be the total output of sector i in some time period. This output can be used either as an input to other sectors (intermediate demand) or consumed as final demand (comprising for example, household consumption, export, government and investment demands).[10] Let $^J X_{ij}$ be the amount of $^J X_i$ used as an input to sector j, and $^J Y_i$ be the amount consumed as final demand. Additionally, each sector purchases inputs in the form of primary factors of production (labour, capital, etc.) and imports from outside the region. Let $^J V_i$ be the amount of these primary inputs used in sector i. The overall input-output balance of this n sector regional economy can now be described in terms of n linear equations:

$$^J X_i = \sum_{j=1}^{n} {}^J X_{ij} + {}^J Y_i \quad (i=1,\ldots,n)$$

(2.3.1)

and represented by the system of accounts depicted in Table 2.1

In order to transform these accounts into an operational model some restrictive assumptions must be made. The main hypothesis is that the input-output structure of any particular sector can be described by a set of technical coefficients $^J a_{ij}$. Each of these coefficients states the level of sector i's output which is required as an input to sector j per unit of sector j's output. That is:

Inputs \ Outputs	Intermediate Sectors				Final Demand Sectors	Total Gross Output
	1	2	...	n		
Intermediate Sectors 1	$^JX_{11}$	$^JX_{12}$	...	$^JX_{1n}$	$^JY_1$	$^JX_1$
2	$^JX_{21}$	$^JX_{22}$	...	$^JX_{2n}$	$^JY_2$	$^JX_2$
.						
.						
.						
n	$^JX_{1n}$	$^JX_{2n}$	...	$^JX_{nn}$	^JY_n	^JX_n
Primary Input Sectors	$^JV_1$	$^JV_2$	...	^JV_n		
Total Gross Outlays	$^JX_1$	$^JX_2$	...	^JX_n		

Table 2.1 A Simplified Regional Transactions Table

$$^J a_{ij} = {^J X_{ij}} / {^J X_j} \qquad (i,j=1,\ldots,n) \tag{2.3.2}$$

It is assumed that these coefficients do not vary over time nor with the level of output in sector j.

Thus, the accounting relation in (2.3.1) can be reformulated as:

$$^J X_i = \sum_{j=1}^{n} {^J a_{ij}} \, {^J X_j} = {^J Y_i} \qquad (i=1,\ldots,n) \tag{2.3.3}$$

which in turn may be rewritten as:

$$\sum_{j=1}^{n} (\delta_{ij} - {^J a_{ij}}) \, {^J X_j} = {^J Y_i} \qquad (i=1,\ldots,n) \tag{2.3.4}$$

where δ is a kronecker delta ($\delta = 1$ if $i = j$, and zero otherwise).

Equation system (2.3.4) can be recognized as a set of simultaneous linear equations in $^J X_i$. Given a set of known or exogenously determined levels of final demand ($^J Y_i$) for region J and a set of technical coefficients ($^J a_{ij}$), then it is possible to solve this system for the levels of production ($^J X_i$) in the endogenous sectors i (i=1,...,n). To see this, the equations in (2.3.4) can be rewritten in matrix notation as:

$$(I - {^J A}) {^J X} = {^J Y} \tag{2.3.5}$$

We can then solve for $^J X$ by premultiplying (2.3.5) by the Leontief inverse $(I - {^J A})^{-1}$. That is:

$$^J X = (I - {^J A})^{-1} \, {^J Y} \tag{2.3.6}$$

Equation (2.3.6) is the usual statement of the input-output model. However, it is possible to extend the model to yield projections not only of output by sector but also of income and employment by sector and population.[11]

However, this flexibility as an operational tool is partially offset by its theoretical weaknesses, since moving from the set of accounting relations in (2.3.1) to the forecasting model in (2.3.6) implies a number of restrictive assumptions. These include:[12]

(a) each commodity is supplied by a single industry or sector, and each sector produces a single commodity (or at most a group of commodities with closely related characteristics);

(b) only one method is used in producing each commodity;

(c) production relations are linear, i.e., inputs purchased by each sector are a fixed proportion of the level of output of that sector.[13] This implies constant returns to scale, and an elasticity of substitution among inputs of zero;

(d) the total effect of carrying out several types of production is the sum of their separate effects. This implies that there are no external economies or diseconomies associated with production;

(e) no capacity constraints exist in any sector, so that the supply of each commodity is perfectly elastic; and

(f) technical coefficients remain relatively stable over time, which ignores such factors as technical change and relative price changes.

These assumptions combine to yield a strange picture of a regional economy; and cause some analysts to dismiss the input-output model as a useless tool of analysis. However, to others these assumptions are acceptable for a sufficient number of important situations so as to warrant extensive use of input-output models provided that appropriate qualifications are made -- and provided that it is recognized that there are situations for which these assumptions are inappropriate and hence for which input-output models can offer no insights, and yet others for which input-output models are useful only when supplemented by other modes of analysis.[14]

Another major problem with the input-output model in (2.3.4) is that it relates to a single region J and requires extension before it could be used to represent a multiregional system. Of course, an input-output model of this type could be constructed for each of the regions comprising the system.[15] However, there are a number of conceptual weaknesses inherent in the use of this set of models for projection purposes. In particular, trade relations between regions make their respective levels of output, income and employment interdependent; and these interdependencies are not captured via the use of separate regional models.

To be more specific, when a regional model is employed an increase in final demand of a given region, say J, will increase the output of industrial sectors of that region by a given amount. But, when an interregional viewpoint is adopted (and trade relations between regions are explicitly recognized) it is found that as output in J increases, inputs are required from other regions -- thus raising the levels of output in those regions. Additional inputs are in turn required from J, thus increasing output in J by more than would have been predicted if these interregional linkages were ignored.[16] Related questions, such as the effect on a given region, say J, of a change in the final demand of another region, say L, cannot be answered via the use of separate regional input-output models.

In the following two sections, a number of interregional input-output models will be presented which are capable of overcoming some of the weaknesses associated with the use of a set of regional input-output models.

2.4 Balanced Regional Input-Output Models

The balanced regional input-output model was originally developed by Leontief (1953) and implemented by Isard (1953). A schematic representation of it is given in Figure 2.1.

To illustrate the workings of the model, let the n sectors of the economy be divided into two mutually exclusive categories:

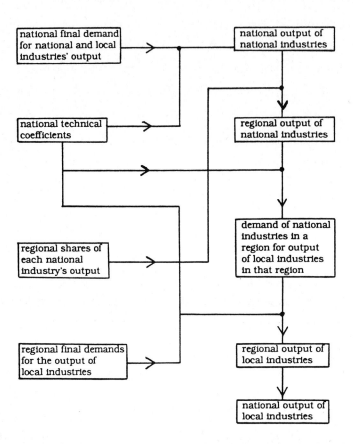

Figure 2.1   A Schematic Representation of the Balanced Regional Input-Output Model

(a) local industries (1,...,k) designated L, which because of transport costs, economies of scale, etc. sell their outputs only in the region in which they are produced; and

(b) national industries (k+1,...,n) designated N, which export their output to other regions besides the one in which they are produced.[17]
The national matrix of input-output coefficients can then be partitioned as follows:

$$A = \begin{bmatrix} A_{LL} & A_{LN} \\ \hline A_{NL} & A_{NN} \end{bmatrix} \quad (2.4.1)$$

where A_{LL} is a k x k submatrix of input-output coefficients indicating levels of production of regionally balanced (local) commodities required per unit output of regionally balanced commodities;

A_{LN} is a k x (n-k) submatrix of coefficients indicating levels of production of regionally balanced commodities required per unit output of nationally balanced commodities;

A_{NL} is a (n-k) x k submatrix of coefficients indicating levels of production of nationally balanced commodities required per unit output of regionally balanced commodities; and

A_{NN} is a (n-k) x (n-k) submatrix of coefficients indicating levels of production of nationally balanced commodities per unit output of nationally balanced commodities.

Given the vector of final demand for the nation as a whole $Y = [\ Y_L\ |\ Y_N\ ]'$, the output of national industries X_N can be determined in typical input-output manner

$$X_N = [I - A]_N^{-1}\ Y \quad (2.4.2)$$

where $[I - A]_N^{-1}$ designates the submatrix of $[I-A]^{-1}$ which relates to direct and indirect requirements for the output of national industries. (See Leontief, 1953).

A basic assumption of the model is that for each national industry the percentage distribution of output among regions remains the same regardless of the level of that output. Let the share of each nationally balanced commodity produced in region J be represented in the diagonal matrix:

$$^J\pi = \begin{bmatrix} ^J\pi_{k+1} & & & \\ & ^J\pi_{k+2} & & \\ & & \cdot & \\ & & & \cdot \\ & & & & ^J\pi_n \end{bmatrix} \quad (2.4.3)$$

where $^J\pi_{k+i} = {^JX_{k+i}}/X_{k+i}$. Region J's output of each nationally balanced commodity can then be determined by multiplying the vector of national output levels by the matrix of share coefficients $^J\pi$. That is:

$$^JX_N = {^J\pi}\, X_N \qquad (J = A,...,U) \quad (2.4.4)$$

Utilizing national input-output coefficients, the inputs of regionally balanced commodities required directly to support the production of these national industry outputs can be determined as $A_{LN}{^JX_N}$. All these inputs must be produced in region J, since by definition no regionally balanced commodity can be transferred from one region to another. In addition to these inputs, region J must provide that part of national final demand for regional commodities, namely ^JY_L, that is consumed in that region.

The sum of these two sets of demands, namely ($A_{LN}{}^J X_N + {}^J Y_L$), is only a first approximation of the required output of regionally balanced commodities in region J. We also need to consider the indirect requirements for regionally balanced commodities. This can be done by premultiplying ($A_{LN}{}^J X_N + {}^J Y_L$) by the inverse matrix $(I - A_{LL})^{-1}$, to yield:

$${}^J X_L = (I - A_{LL})^{-1} (A_{LN}{}^J X_N + {}^J Y_L) \qquad (2.4.5)$$

as the required output of each local industry in region J. In a similar fashion, every other region's output of national and local industries can be determined.

Recent applications of this type of model include the IDIOM model (Dresch, 1980)[18] and the ORANI-ORES model (Madden, et al. 1981, 1982)[19].

In addition to the shortcomings that generally characterize input-output models, there are others that are peculiar to the balanced regional model. Namely:

(a) the classification of industries into the nationally balanced and regionally balanced categories involves both technical and conceptual difficulties which are difficult to resolve (see Isard, 1953);

(b) the assumption that the ${}^J\pi$ coefficients are constant is unrealistic. There is no reason to believe that as final demand changes all regions will expand or contract output in any national industry in fixed proportion;

(c) the use of national input-output coefficients to characterize interindustry relations in each region is unrealistic, since it precludes differences in production patterns among regions;

(d) when the household sector is included in the structural matrix, the use of national input-output coefficients is even more unrealistic, since it also precludes differences in consumption patterns among regions;

(e) the model does not deal with interregional relationships, and so is unable to incorporate interregional feedback effects into its impact projections;

(f) while the model is particularly useful for examining the regional implications of national projections or policies, its top-down character makes it much less useful for examining the national implications of regional projections or policies. Indeed, some major modifications to its basic structure are required in order to perform the latter types of analyses (See Isard, 1953).

2.5 Interregional Input-Output Models
2.5.1 Isard's Ideal Type

The original interregional model was formulated by Isard (1951). To illustrate the workings of the model, consider a multiregion system comprising U regions (denoted J,L=A,...,U). Let there be the same n sectors (denoted i,j=1,...,n) in each region. The distinctive feature of this 'ideal' interregional model is that it treats these apparently identical sectors in each region as distinct separate industries. Thus a U region, n industry model becomes equivalent to an input-output table containing Un sectors.

The basic set of structural and flow relationships can be described in terms of Un linear equations of the form:

$$^J X_i = \sum_{L=A}^{U} \sum_{j=1}^{n} {}^{JL}X_{ij} + {}^J Y_i \quad \begin{matrix} (i=1,...,n) \\ (J,L=A,...,U) \end{matrix}$$

(2.5.1)

where $^{JL}X_{ij}$ = flow of output from sector i in region J to satisfy intermediate input demand from sector j in region L; and

$^J Y_i$ = flow of output from sector i in region J to satisfy final demand (regardless of region of origin of these demands).

As in the single region case discussed in section 2.3, in order to transform the accounting relation given by (2.5.1) into an operational model we need to employ some restrictive assumptions (the implications of which will be discussed later in this section). In this case, the main hypothesis is that the input-output structure of a particular sector j in any individual region L can be described by a set of interregional 'technical-trade' coefficients. The typical element $^{JL}\hat{a}_{ij}$ in this set states the amount of a particular input from sector i in region J which is required by sector j per unit of its own output in region L. That is:

$$^{JL}\hat{a}_{ij} = {}^{JL}X_{ij} / {}^{L}X_j \quad \begin{array}{l}(i,j=1,\ldots,n)\\(J,L=A,\ldots,U)\end{array} \quad (2.5.2)$$

where it is assumed that these coefficients do not vary over time or with the level of output of sector j in region L. Thus, the structural and flow relationships in (2.5.1) can be reformulated as:

$$^{J}X_i = \sum_{L=A}^{U} \sum_{j=1}^{n} {}^{JL}\hat{a}_{ij} \, {}^{J}X_j + {}^{J}Y_i \quad \begin{array}{l}(i=1,\ldots,n)\\(J,L=A,\ldots,U)\end{array} \quad (2.5.3)$$

which can then be rewritten as:

$$({}^{JL}\delta_{ij} - {}^{JL}\hat{a}_{ij}) = {}^{J}Y_i \quad \begin{array}{l}(i=1,\ldots,n)\\(J,L=A,\ldots,U)\end{array} \quad (2.5.4)$$

where $^{JL}\delta_{ij}$ is a kronecker delta ($^{JL}\delta_{ij} = 1$ if $i \neq j$ and $J \neq L$, and 0 otherwise).

Equation system (2.5.4) represents a set of Un simultaneous equations in $^{J}X_i$, where $J = A,\ldots,U$ and $i=1,\ldots,n$. Given a set of known or exogenously determined levels of final demands ($^{J}Y_i$) for each region, and a set of interregional 'technical-trade' coefficients ($^{JL}\hat{a}_{ij}$), then it is possible to solve this system for the levels of production ($^{J}X_i$) in each region and sector. To see this, the equations represented in (2.5.4) can be rewritten in matrix notation as:

$$(I - Æ) X = Y \qquad (2.5.5)$$

We can then solve for X by premultiplying (2.5.5) by the Leontief-type inverse $(I - Æ)^{-1}$. That is:[20]

$$X = (I - Æ)^{-1} Y \qquad (2.5.6)$$

Hartwick (1971) has shown that the interregional 'technical-trade' coefficients in equation (2.5.2) can be partitioned into two components. That is:

$$(^{JL}\hat{a}_{ij}) = (^{JL}c_{ij})(^{L}a_{ij}) = \frac{^{JL}X_{ij}}{^{L}X_j} \frac{^{L}X_{ij}}{^{L}X_j} \qquad \begin{array}{l}(i,j=1,\ldots,n)\\(J,L=A,\ldots,U)\end{array} \qquad (2.5.7)$$

where $^{JL}c_{ij}$ = a trade coefficient representing the proportion of total purchases of commodity i by industry j in region L which is produced in region J;[21] and

$^{L}a_{ij}$ = a technical coefficient defined in a manner identical to those in the single region model.

Equation (2.5.6) can then be rewritten, following Polenske (1980a), as:

$$X = (I - CA)^{-1} Y \qquad (2.5.8)$$

where C = a nU x nU square matrix of expanded trade coefficient matrices, with the trade coefficients arrayed along the principal diagonal of the UxU blocks and zeros for off-diagonal elements; and

A = a nU x nU block diagonal square matrix, with the technical coefficients for each region appearing as U square matrices (nxn) on the diagonal blocks and zeros appearing in the off-diagonal blocks.

This reformulation will prove useful later when comparing the model with others that have been proposed for use in interregional analysis.

In order to derive estimates of the $^{JL}\hat{a}_{ij}$'s however, one needs access to interregional trade flows by region of origin and destination, and by sector of production and use. Such detailed trade statistics are not generally available from published sources, and the costs of generating them from surveys are prohibitive for any reasonable level of sectoral and spatial disaggregation. As a result, this ideal model has been implemented in only a limited number of cases.[22] Nevertheless, it is generally regarded as a useful reference norm from which other interregional models can be evaluated.

For example, the closest approximation to this model in the United States is the one developed by Riefler and Tiebout (1970) for the states of California and Washington. Survey-based input-output tables were available for both the California and Washington economies (see Hansen and Tiebout (1963) and Bourque et al (1966), respectively) which permitted the estimation of intraregional coefficients $^{WW}\hat{a}_{ij}$ and $^{CC}\hat{a}_{mk}$ (where W denotes Washington, C denotes California, i,j =1,...,31 represents Washington producing and consuming industries, respectively, and m,k =1,...,22 represents California producing and consuming industries, respectively).

In the case of the Washington economy, an import matrix was available, the typical element of which specified the proportion of sector j's input requirements for commodity m which are imported from the 'rest of the world'. By assuming a constant proportion of Washington's imports of commodity m came from California regardless of which sector it was eventually used in, it was possible to estimate the interregional coefficients $^{CW}\hat{a}_{mj}$. These California to Washington shares were estimated from published trade flow statistics. In the case of the California economy, the interregional coefficients were estimated indirectly from an export matrix that was available for Washington. The typical element of this matrix specified the proportion of sector i's output which was exported to the 'rest of the world' for use in the production of commodity k. By assuming a constant proportion of Washington's exports of commodity i went to California regardless of which sector it was eventually used in, it was possible to estimate the $^{WC}\hat{a}_{ik}$ coefficients. These Washington to California shares were once again estimated from published trade flow statistics.

The basic interregional 'technical-trade' coefficients matrix was then used in the manner outlined above to project the direct and indirect output requirements by sector and region for a given set of final demands.

Even when data availability and research resources permit a better approximation to the 'ideal' given in (2.5.6), the assumptions that the $^{JL}\hat{a}_{ij}$'s are constant over time and independent of changes in the level of output represent a major drawback of the model. This is so since they require:

(a) constant intraregional technical coefficients (i.e., the $^{L}a_{ij}$'s in equation 2.5.7). The implications of this requirement have already been outlined in section 2.3 when discussing the single region model and will not be repeated here; and

(b) stable interregional trade coefficients (i.e., the $^{JL}c_{ij}$'s in equation 2.5.7). Since trading patterns reflect regional cost-price relationships, rather than technological requirements, this implies:

-constancy of regional production costs;

-constancy of cost of transporting a unit of a good between each pair of regions;

-'infinitely' elastic factor supply curves around existing factor prices in each region;

-'unlimited' production capacity in each industry in each region;

-'unlimited' capacity in transport network between every pair of regions.[23]

Once again these assumptions combine to yield a strange picture of a multiregional economy. However, the 'ideal' interregional input-output model is potentially a very useful tool of analysis, especially if supplemented with other interregional models in such a way as to eliminate some of these highly restrictive assumptions.

2.5.2 The 'Fixed Shares of Total Purchases from Each Region' Type

While an interregional model of the Isard-type requires commodity flow data by both region and sector of origin and destination, the commodity flow data available in many countries specify only the region of origin and destination, but not the sector of destination. The 'fixed shares of total purchases from each region' variant (Chenery, 1953 and Moses, 1955) is tailored to suit this situation.

To illustrate the workings of the model, consider again the case of a U region system with n identical sectors in each region. The balance equations of the interregional system are given by:

$$^J X_i = \sum_{j=1}^{n} \sum_{L=A}^{U} {}^{JL}X_{ij} + \sum_{L=A}^{U} {}^{JL}Y_i \quad \begin{array}{l}(i=1,...,n)\\(J=A,...,U)\end{array} \quad (2.5.9)$$

where $^{JL}Y_i$ = flow of output from sector i in region J to satisfy final demand in region L.

This set of Un equations differs from those in equation (2.5.1) only by its identification of the region of origin of the final demands $^L Y_i$. This system too can only be solved by introducing some additional restrictions. In this case, the restrictions imposed take the form of two sets of coefficients.

The first set of coefficients describe the input-output structure of each sector in each region, in a manner analogous to the single region model discussed in section 2.3. The typical element in this set, namely $^J a_{ij}$ states the input of commodity i required by sector j from all regions per unit of its own output in region J. Once again, these technical coefficients are assumed to remain constant over time and to be independent of change in the level of output to sector j in region J.

The second set of coefficients describe the trade pattern between regions for each commodity. The typical element in this set, namely $^{JL}c_i$ states the proportion of region L's requirements for commodity i which is purchased from region J.[24] That is:

$$^{JL}c_i = (^{JL}X_i) / (\sum_j {}^L X_{ij} + {}^L Y_i) \quad \begin{matrix}(i=1,\ldots,n)\\(J,L=A,\ldots,U)\end{matrix} \quad (2.5.10)$$

where $^{JL}X_i$ = flow of commodity i from region J to region L;[25]

$^L Y_i$ = amount of commodity i consumed by final users in region L regardless of where it was produced.

It is assumed that these proportions are identical for each receiving industry. For example, if region L imports γ percent of its requirements of commodity i from region J, then each industry j in region L imports δ percent of its requirements of i from L. That is:

$$^{JL}c_{i1} = \ldots = {}^{JL}c_{ij} = \ldots = {}^{JL}c_{in} = {}^{JL}c_i \qquad (2.5.11)$$

(The implications of this assumption will be discussed later in this section). This set of trade coefficients is also assumed to remain constant over time.

With the use of these two sets of coefficients, the balance equations in (2.5.9) can be rewritten as:

$$^J X_i = \sum_{L=A}^{U} (^{JL}c_i) \left[\sum_{j=1}^{n} (^L a_{ij}) {}^L X_j + {}^L Y_i \right] \qquad (2.5.12)$$

or alternatively as:[26]

$$^J X_i = \sum_{L=A}^{U} \sum_{j=1}^{n} (^{JL}c_i)(^L a_{ij}) {}^L X_j + \sum_{L=A}^{U} (^{JL}c_i) {}^L Y_i \qquad (2.5.13)$$

Given estimates of the technical coefficients ($^L a_{ij}$) for each region, and trade coefficients ($^{JL} c_i$) for each commodity, it is now possible to solve the system of equations given as (2.5.9) for the set of production levels ($^J X_i$) in each region corresponding to any given set of final demands. To see this equation (2.5.13) is rewritten in matrix form as:

$$X = C(AX + Y) \tag{2.5.14}$$

where X = a column vector (nU x 1) giving the level of output in each region and sector, with a typical element $^J X_i$ defined as above;

Y = a column vector (nU x 1) giving the level of final demands for each commodity in each region, with a typical element $^J Y_i$ defined as above;

A = a (nU x nU) block diagonal matrix with U square matricies (n x n) of input coefficients $^J a_{ij}$ along the main diagonal;

C = a (Un x Un) square matrix containing n diagonal matricies (U x U) with a typical element $^{LJ} c_i$ defined as above.

Equation (2.5.14) can then be rewritten as:

$$(I - CA) X = CY \tag{2.5.15}$$

which can be solved by premultiplying by the inverse $(I - CA)^{-1}$ to yield:[27]

$$X = (I - CA)^{-1} CY \tag{2.5.16}$$

This type of model was applied by Chenery (1954) to a study of the Italian economy. The economy was divided into two regions (North and South) and 23 sectors. Each region's technical coefficients matrix $^J A$ (excluding households) was estimated to be identical to the national A matrix when the latter was aggregated to 22 industrial sectors.[28] In deriving the trade coefficients matrix for each commodity, the 23 sectors were divided into 3 subclasses:

(a) national -- 10 sectors for which supply and demand were assumed to balance only at the national level, and for which the relative shares of local, interregional and foreign sources of supply were assumed to be identical for both regions;

(b) local -- 9 sectors for which supply and demand were assumed to balance at the local level in both regions, and for which no interregional trade and only a small (percentage equal in both regions) level of foreign trade was assumed to take place; and

(c) mixed -- remaining 4 sectors (including households) for which it was assumed: (i) the South did not import from the North, and hence derived all its supply from local and foreign sources in fixed proportions; and (ii) the North drew on all three sources of supply in fixed proportions.

The model was then used to examine the effects of investment programs carried out in the South on household income in both regions. The results revealed that the use of a regional rather than an interregional model (that is disregarding the interregional effect) resulted in an increase of 18 percent in household income projections for the Southern region.

At about the same time, Moses (1955) applied the model to a study of the United States economy. The economy was divided into 3 regions (East, Midwest and West) and 11 sectors (including households). Once again the national input-output coefficients were used in each region. The derivation of trade coefficients presented more problems in this case, since it could no longer be assumed that an export from one region was an import of the other. Direct use of interregional commodity flow data was required, and where this data was not available from published sources it was approximated via the use of production and consumption data and 'professional judgement based on locational considerations'. The model was then used to examine the effects of an increase in non-consumption final demands in the East on production levels in each region.

Although both these applications assumed that each region's technical coefficients were identical to the corresponding national coefficients, this was dictated by the paucity of available data at that time rather than as an essential feature of operational versions of the model. For example, as discussed in part 5 of this section, Polenske (1972, 1980) has been able to implement a model for the United States which employs regional-specific input-output coefficients.

However, while Isard's 'ideal' interregional input-output model has been transformed into a form that can more easily be made operational, this has (as outlined above) necessitated the introduction of yet another highly restrictive assumption -- namely, that trade patterns by type of input are not only constant at an aggregate regional level, but also constant and identical for each industry using that particular input in the given region. Chenery (1953) rationalizes the latter part of this assumption by arguing that all sectors in a region requiring an input can be regarded as constituting a single market for it, and sources of supply (local, interregional, foreign) are determined more by the level of demand in this market than by the nature of intended use. Nevertheless, the assumption represents a gross simplification of reality for many (if not most) commodities, and should be borne in mind when the results of the model are used in a policy context.

2.5.3 The Fixed Shares of Total Sales to Each Region' Type

Polenske (1970) tested another type of interregional input-output model which requires the same amount of basic data as the Chenery-Moses model just discussed. She dubbed this alternative the 'fixed row coefficient' model. The essential difference between the two models stems from the set of coefficients employed to describe the commodity trade relationships between regions.

While the Chenery-Moses trade coefficients ($^{JL}c_i$) are defined by allocating a given region's total purchases of a commodity between regions in fixed proportions, the trade coefficients in this model (denoted $^{JL}r_i$) are defined by allocating a given region's total sales of a commodity between regions in fixed proportions.

Thus the typical element $^{JL}r_i$ in the fixed row coefficient model's set of trade coefficients states the proportion of total production of commodity i in region J which is sold to any region L. It is defined as:[29]

$$^{JL}r_i = {^{JL}X_i}/{^{J}X_i} \qquad (J,L=A,...,U) \qquad (2.5.17)$$

and is assumed constant over time and with changes in the level of $^{J}X_i$.

With the use of the $^{JL}r_i$ coefficients instead of the $^{JL}c_i$ coefficients, equation (2.5.14) becomes:

$$R'X = AX + Y \qquad (2.5.18)$$

where R   = a (Un x Un) square matrix containing n diagonal matricies (U x U) with a typical element $^{JL}r_i$, and

R' = a (nU x nU) square matrix formed by interchanging the rows and columns of R.

Equation (2.5.18) can then be rewritten as:

$$X = (R' - A)^{-1} Y \qquad (2.5.19)$$

which can be used to estimate the set of output levels by sector and region corresponding to any given set of final demands.

However, use of the $^{JL}r_i$ coefficients to represent trade relations between regions is highly suspect on theoretical grounds -- it implies, for example, that the proportion of region J's total production of commodity i that is sold to region L is constant irrespective of any change in the level of demand for it in L or any other region.

Polenske (1970), conducting tests using Japanese data, found that the performance of the 'fixed shares of total sales to each region' type of model was consistently worse than that of the 'fixed shares of total purchases from each region' type of model.[30] This suggests that regional purchase patterns are, on average, more stable than regional sales patterns. However, more testing on the relative performances of the two models is required before any definite conclusions can be made concerning the relative merits of the two models.

2.5.4 Leontief-Strout and Other Gravity-Type Models

An alternative approach to modelling multiregional input-output relationships has been proposed by Wilson (1970). The feature which distinguishes this family of models from those discussed above is their linkage of a set of separate regional input-output models via the use of a set of gravity-type equations representing interregional trade relationships.

The basic assumptions underlying the Leontief-Strout variant are that the ultimate destination of their output is irrelevant to producers, and that the origin of a particular type of commodity is irrelevant to its users. (Leontief and Strout, 1963). This implies that all producers of commodity i in region J can be visualized as passing their output to a single regional supply pool, while all users of commodity i in region L can be visualized as ordering and receiving their requirements from a single regional demand pool. All interregional movements of commodity i can then be considered to be shipments from regional supply to regional demand pools of that commodity. A schematic representation of the relevant flows in such a system is given in Figure 2.2.

To illustrate the workings of the model, once again assume a system comprising U regions, each with n identical sectors. The basic input-output relations for each region J can be represented as:

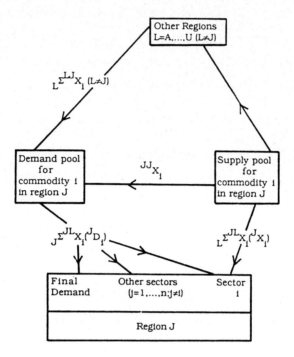

(After Wilson, 1970, p53)

Figure 2.2 Schematic Representation of Flows of Commodity i in Leontief-Strout Gravity Model

$$^JD_i = \sum_{j=1}^{n} {}^Ja_{ij}\,{}^JX_j + {}^JY_i \quad \begin{matrix}(i=1,\ldots,n)\\(J=A,\ldots,U)\end{matrix} \tag{2.5.20}$$

where ^JD_i is the total demand for commodity i in region J, and is comprised of two components, namely demand by intermediate users ($\Sigma\,{}^Ja_{ij}\,{}^JX_j$) and final users ($^JY_i$).[31]

In addition, the interregional flows of each commodity i must satisfy the following conditions.[32]

$$^JX_i = \sum_{L=A}^{U} {}^{JL}X_i \quad \begin{matrix}(i=1,\ldots,n)\\(J,L=A,\ldots,U)\end{matrix} \tag{2.5.21}$$

$$^LD_i = \sum_{J=A}^{U} {}^{JL}X_i \quad \begin{matrix}(i=1,\ldots,n)\\(J,L=A,\ldots,U)\end{matrix} \tag{2.5.22}$$

$$X_i = \sum_{J=A}^{U} {}^JX_i = \sum_{L=A}^{U} {}^LD_i = D_i \quad (i=1,\ldots,n) \tag{2.5.23}$$

Equation (2.5.21) shows that a balance must exist between total production of commodity i in region J and the total amount of i shipped from J to all regions (including itself); while equation (2.5.22) shows that a balance must exist between total demand (usage) of commodity i in region L and the total amount of i shipped to L from all other regions (including itself). Thus ^JX_i can be regarded as the supply pool of i in J, ^LD_i as the demand pool for i in L, and $^{JL}X_i$ the total shipment of i from its supply pool in J to the demand pool for i in L. Equation (2.5.23) shows that, when summed over all regions, a balance must exist between aggregate supply and aggregate demand for each commodity i.

The set of equations used to 'explain' interregional flows of each commodity i are of the form:

$$^{JL}X_i = \frac{^{J}X_i \, ^{L}D_i \, ^{JL}q_i}{X_i} \tag{2.5.24}$$

or equivalently

$$^{JL}X_i = \frac{^{J}X_i \, ^{L}D_i \, ^{JL}q_i}{D_i} \tag{2.5.25}$$

where $^{JL}q_i$ = an empirical constant (to be discussed in more detail later) which reflects the cost of transporting commodity i from J to L.

Equation (2.5.24) can be interpreted as stating that the share of L's total demand for commodity i that is supplied by J, namely ($^{JL}X_i/^{L}D_i$), is proportional to the share of aggregate supply of commodity i that comes from production in J, namely ($^{J}X_i/X_i$), after this share has been weighted by a parameter, namely ($^{JL}q_i$), reflecting the cost of transportating i from J to L. Similarly, equation (2.5.25) can be interpreted as stating that the share of J's total production of commodity i that is supplied to L, namely ($^{JL}X_i/^{J}X_i$), is proportional to the share of aggregate demand for commodity i that comes from users in L, namely ($^{L}D_i/D_i$), after this share has been weighted by $^{JL}q_i$.

Since the $^{JL}q_i$ parameters are treated as constants, this implies that the cost of transporting i from J to L (a) is constant over time, (b) does not change with the volume of i shipped from J to L, and (c) is not influenced by the volume of other commodities j shipped from J to L. This in turn implies 'unlimited' capacity on the transport network between J and L, and constant interregional transport costs.

Providing neither $^{J}X_i$ nor $^{L}D_i$ is equal to zero, and if $^{JL}q_i$ is positive, then equations (2.5.24) and (2.5.25) will predict a positive flow $^{JL}X_i$ from J to L. On the one hand, this creates a problem because the model will predict many more non-zero shipments than actually occur.[33] However, this property does permit the model to capture (at least to some extent) the reality of cross-hauling -- that is, simultaneous flows of i in opposite directions between a pair of regions.[34]

By substituting the interregional flow equation (2.5.24) into equations (2.5.21) and (2.5.22) we obtain:[35]

$$^{J}X_i = {}^{JJ}X_i + \frac{{}^{J}X_i \sum_{L=A}^{U} ({}^{L}D_i \, {}^{JL}q_i)}{X_i} \qquad ({}^{JJ}q_i=0; J=A,\ldots,U; i=1,\ldots,n)$$

(2.5.26)

and

$$^{L}D_i = {}^{LL}X_i + \frac{{}^{L}D_i \sum_{J=A}^{U} ({}^{J}X_i \, {}^{JL}q_i)}{D_i} \qquad ({}^{LL}q_i=0; L=A;,,,U; i=1,\ldots n)$$

(2.5.27)

Equation (2.5.26) shows that the total production of i in J is equal to the amount produced and sold to itself plus the amount exported to other regions; while equation (2.5.27) shows that total usage of i in L is equal to the amount produced and used in that region plus the amount imported from other regions.

This completes the basic system of equations, however, we can express the system in a form which makes computation easier by eliminating the intraregional flows ($^{JJ}X_i$ and $^{LL}X_i$). To do this equation (2.5.27) is rewritten by substituting J for L to obtain:

$$^{J}D_i = {}^{JJ}X_i + \frac{{}^{J}D_i \sum_{L=A}^{U} ({}^{L}X_i {}^{LJ}q_i)}{X_i} \quad ({}^{JJ}q_i=0; J=A,\ldots,U; i=1,\ldots,n)$$

(2.5.28)

Then combining equations (2.5.26) and (2.5.28) yields:

$$^{J}D_i X_i - {}^{J}D_i \sum_{L=A}^{U} ({}^{L}X_i {}^{LJ}q_i) = {}^{J}X_i X_i - {}^{J}X_i \sum_{L=A}^{U} ({}^{L}D_i {}^{JL}q_i)$$

(2.5.29)

where $J = A, \ldots, U$, and $^{JJ}q_i = {}^{LL}q_i = 0$.

Assuming that the required set of final demands ($^{J}Y_i$), technical coefficients (^{J}A) and trade coefficients ($^{JL}q_i$) are given, then a general solution to the above set of equations (i.e., (2.5.21), (2.5.22), (2.5.23), and (2.5.29)) can be found which specifies the output levels for each industry in each region. However, the interregional flow equations in (2.5.29) are non-linear. Thus for the purpose of numerical solution, they have to be linearized by means of a first order approximation. This involves splitting each variable into two components: its base year magnitude and an incremental component. Assuming that base year magnitudes are known for all variables, the model can then be used to predict increments of all endogenous variables (i.e., $\Delta^{J}X_i$ and $\Delta^{J}D_i$) given a set of increments of regional final demands (i.e., $\Delta^{J}Y_i$).[36]

One more question needs to be resolved, namely how does one go about estimating the trade coefficients $^{JL}q_i$? Firstly, if base year trade flows $^{JL}X_i$ are available for all commodities and pairs of regions then the corresponding $^{JL}q_i$ can be estimated from (2.5.24). Assuming this $^{JL}q_i$ remains constant, it can be used to estimate trade flows in subsequent periods. This is called the point-estimate version of the model. Otherwise, Leontief and Strout (1963) suggest that they be estimated as:

$$^{JL}q_i = (^JC_i + {}^LK_i) {}^{JL}d_i {}^{JL}\delta_i \qquad (2.5.30)$$

where ^JC_i and ^LK_i are parameters reflecting what is termed the 'relative position' of J as a supplier and L as a user of commodity i, respectively;

$^{JL}d_i$ is the inverse of transport cost for i between J and L (or where this information is not available, simply distance between J and L); and

$^{JL}\delta_i$ is a kroneker delta taking the value of 0 when it is known that $^{JL}x_i = 0$, or when J=L, and 1 otherwise.

The problem then becomes how to estimate the ^JC_i and ^LK_i parameters, since they cannot be observed. Two alternative methods were suggested by Leontief and Strout (1963); namely:

(a) application of least squares or other curve-fitting procedure to equation (2.5.24) when $(^JC_i + {}^LK_i) {}^{JL}d_i\delta_i{}^{JL}$ replaces $^{JL}q_i$. This is called the least squares version of the model.

(b) solution of the set of simultaneous equations including (2.5.25), (2.5.26), (2.5.27), and (2.5.30) using base year magnitudes for ^Jx_i, ^LD_i, and $^{JJ}x_i$. This is called the exact solution of the model.

However, regardless of what method is used to derive ^JC_i and ^LK_i, the assumption that they remain constant seems unrealistic.

The model has been applied by Leontief and Strout (1963) and Polenske (1970, 1972). Leontief and Strout used the model to predict rail shipments of coal, cement, soybean oil and steel shapes for 9-13 United States' regions. Polenske's studies involved the use of the 1960 Japanese interregional input-output table to derive base year estimates for 10 sectors and 9 regions. In these studies she (a) compared the relative performances of the different gravity model formulations outlined above, and concluded that although the point estimate version requires the most data it usually produces the lowest errors of estimation; and (b) compared the relative performances of the Leontief-Strout model (point estimate version), the Chenery-Moses model, and the fixed row coefficient model. While the Leontief-Strout model almost invariably produced lower errors of estimation than the fixed row coefficient model, she found no discernible difference in the overall predictive ability of the Chenery-Moses and Leontief-Strout models.

Wilson (1970) shows how more general gravity models could be used to replace equation (2.5.24) and how the resulting model can be solved via entropy maximizing principles.[37]

2.5.5 Polenske's MRIO Project

A multiregional input-output model (MRIO) that includes 44 regions, 78 industries and 6 final demand categories has been implemented for the United States by Polenske (1972), (1980). In order to do this, three major sets of data had to be assembled (estimated) employing a consistent region and industrial classification. These data comprised:

(a) intraregional interindustry flows for each region;

(b) interregional trade flow data for each commodity subject to data availability; and

(c) base year final demand by sector by region.[38]

Since interregional trade flow data could only be estimated in the form of $^{JL}X_i$ not $^{JL}X_{ij}$, the Isard 'ideal' interregional model could not be adopted. Instead, Polenske conducted experiments with both: the 'fixed shares of total purchases from each region' model, and the Leontief-Strout gravity model (point estimate version). However, problems were encountered in trying to get a convergent solution to the Leontief-Strout gravity model. Therefore, the 'fixed shares of total purchases from each region' model was the one finally adopted.[39]

The MRIO model has subsequently been employed by Polenske (1972) to derive estimates of regional output and interregional trade flows corresponding to given projections of regional final demands; and by others (see Polenske (1980)):

(a) to project the regional employment and income effects of consolidation of the northeast railroads into what is now called CONRAIL;

(b) to project employment in Massachusetts to 1980 and to obtain output, income and employment multipliers;

(c) together with a microsimulation model of household behavior and a macrosimulation model of business behavior, to estimate the regional, industrial and income distributional effects of a national negative income tax and family assistance program; and

(d) together with emission coefficients and abatement cost coefficients to estimate the regional price, output and interregional trade impacts of investment in the pollution abatement industry.

Because of its high level of regional and sectoral disaggregation, and its use of a different set of intraregional technical coefficients for each region, the implementation of this model represented a major advance in the state of the art in multiregional input-output modelling in the United States. Whether a comparable feat could be performed in other countries depends on:

(a) the resources (financial and manpower) available for regional research; and

(b) the nature of data (published and unpublished) available, or likely to become available, by industry and region.

2.6 Concluding Remarks

This chapter has dealt with the core component of the integrated multiregion model -- namely that which determines (projects) the level of economic activity in each region.

The economic base approach was considered first, but in general it was dismissed as inferior should the resources exist to support the operation of an input-output module. A number of variants of the input-output approach were outlined, and the conclusion drawn that an interregional input-output table was the most appropriate for use in an integrated model. For details of the approach to be adopted for deriving such a table in the Australian context, see Smith (1983, 52-164). This approach involves the use of a mixture of the approaches outlined in sections 2.5.1-2.5.5, as required by available data and/or as suggested by basic economic theory for each sector in each region. For example, for some sectors (eg. mining and mineral processing) use is made of Isard's ideal method, for other sectors (eg. some manufacturing sectors) use is made of the fixed column coefficient approach, while for other sectors (eg. some manufacturing sectors) use is made of a variant of the Leontief-Strout approach.

NOTES

[1] See Hoyt (1949) and Tiebout (1962) for some early applications.

[2] This is only one variant of the economic base multiplier. For a discussion of others, see Isard and Czamanski (1965).

[3] For example, categorization of each activity as either wholly basic or wholly non-basic is naive; firm-by-firm surveys are costly and time-consuming; simple location quotient methods assume each region's pattern of productivity, resource use and per capita consumption are the same as the national average, and that there are no net exports nationally; and the minimum requirements approach begs the question of how to determine a region's minimum requirements for each activity. Among others, see Isard et al (1960), Isserman (1977), Pratt (1968) and Moore (1975) for further discussion.

[4] For example, the model can be calibrated by (a) using one observation, in which case projections relate to levels of economic activity; (b) using two observations, in which case projections relate to changes in activity; or (c) using a time series of observations, in which case different lag structures can be incorporated into projections. For further discussion, see Isard and Czamanski (1965), Sasaki (1963), Part (1970) and Moody and Puffer (1970).

[5] Generally speaking, the greater the level of sectoral disaggregation, the lower the mulitplier estimate. Among others, see Greytak (1969) and Isserman (1977).

[6] Tiebout (1962) suggested that the analyst should act to eliminate the first of these multiplier components by treating any activity linked directly to basic activity by interindustry sales as part of basic activity. Isard and Czamanski (1965) go further, and propose a model which permits the multiplier to be disaggregated into these two component parts.

[7] In an attempt to eliminate this latter source of criticism, a number of analysts have adopted a differential multiplier approach in which the basic sector is disaggregated into several sectors. (See Weiss and Gooding (1968), Garnick (1970), Brownrigg and Green (1975)). However, although this disaggregation improves the model and makes even clearer its status as a member of the same family as input-output models (see Isard and Czamanski (1965), Billings (1969), Garnick (1969)), it seems even more general to employ a full-fledged input-output model since this permits projection of the sectoral distribution of impacts as well as providing an aggregate impact estimate.

[8] The model seems to imply either that each region has sufficient excess capacity that an expansion of the basic sector does not require resource shifts out of the non-basic sector, or that resources are perfectly mobile between regions and hence can respond "automatically" to changes in the distribution of basic activity.

[9] Although the size of the leakage from each region's economy, and hence the size of its multiplier ^Jk, is dependent on the level of imports, the fact that one region's exports comprise another region's imports is not explicitly recognized.

[10] The sectors included as components of final demand will vary with the problem at hand, the region under study, data availability and many other factors. For example, while the household sector has traditionally been classified as a component of final demand at the national level, it is frequently treated as an endogenous 'production' sector in regional applications. When this is done, households are viewed as selling labor (and other services) as 'output' and using the income (primarily wages and salaries) received in return to purchase 'inputs' of food, clothing and other consumption goods.

[11] Extension to yield income multipliers involves incorporating the household sector within the structural ^JA matrix; extension to yield employment multipliers involves the use of a set of sectoral output to employment coefficients, and extension to yield population estimates involves the use of both output to employment coefficients and employment to population coefficients. For detailed discussion, see Richardson (1972), and Jensen et al (1977) among others.

[12] For further discussion, see Isard et al (1960), Chenery and Clark (1959) and Miernyk (1965).

[13] When households are included in the ^JA matrix, consumption relations are also assumed to be linear, i.e., consumption goods purchased by the household sector are a fixed proportion of the level of its income.

[14] See Isard et al (1960) for further elaboration of this type of philosophy.

[15] See Jensen et al (1977) for one of the best examples of the use of such a set of models.

[16] A number of attempts have been made to quantify the extent of error associated with ignoring these feedback effects. For example, see Miller (1966), Greytak (1970) and Gillen and Guccione (1980). Not surprisingly, the size of the error has been found to depend on the size of the regions comprising the system, and the degree of self-sufficiency of the respective regional economies. This suggests that if the multiregion

system under study comprises a set of regions which trade among themselves to only a limited extent, then the errors associated with the use of a set of regional models may not be too severe.

[17] The model can be expanded to embrace more than two orders of regions and two classes of industries, however the basic structure remains the same. See Leontief (1953) and Isard et al (1960) for further discussion.

[18] This United States' model uses states as its regions, and operates with 86 industries of which 60 are treated as nationally balanced.

[19] This Australian model operates with 113 sectors of which 29 are treated as regionally balanced, and uses states as its regions.

[20] Note that this solution has the same form as the single region model (see equation 2.3.6), except that here the final demand vector Y (consisting of Un elements) has replaced the vector ^JY (consisting of n elements), the output vector X (consisting of Un elements) has replaced the vector ^JX (consisting of n elements), and the A matrix (consisting of Un x Un elements) has replaced the ^JA matrix (consisting of n x n elements). Also, as in the single region case, this model is capable of extension to yield projections of income by sector by region, employment by sector by region and population by region.

[21] Note that by definition $\sum_J {}^{JL}c_i = 1$.

[22] For example, the Japanese government has published a 9 region, 11 industry interregional input-output model for Japan. See MITI (undated).

[23] In any given application the last three requirements can be relaxed somewhat. What is then required is that to satisfy the postulated level of final demands no region should be predicted to supply any more of any good than its factor endowment and available production capacity can produce, and the level of commodity flows predicted by the model does not exceed the capacity of the transport network between any pair of regions.

[24] The sum $\sum_J {}^{JL}c_i$ must equal one, since the coefficients are proportions of total requirements.

[25] If interregional trade flow statistics of the form $(^{JL}X_i)$ are not available, then they can be estimated by indirect techniques. However, each of these indirect methods has major defects. For example, most relate to net rather than gross flows, yet cross-hauling may be very

important -- especially at the level of sectoral aggregation used in most interregional interindustry studies.

[26] Comparing the Chenery-Moses coefficient ($^{JL}c_i$) ($^L a_{ij}$) with the Isard coefficient ($^{JL}c_{ij}$) ($^L a_{ij}$), defined by equation (2.5.7), reveals very clearly the essential difference between the two models. That is while the Chenery-Moses model assumes that the sources of supply are fixed for all uses of a given commodity in a given region, the Isard model permits the source of supply to depend on the type of use to which the commodity is to be put. Thus there is an explicit averaging involved in the Chenery-Moses variant which is not found in the more precise Isard variant. See Hartwick (1971) for a more detailed comparison of the two models.

[27] This equation is sometimes written in the form $X = (C^{-1} - A)^{-1} Y$, since this reduces the computer capacity required to solve the system. See Polenske (1980).

[28] When the household sector was made endogenous, somewhat different coefficients were employed in each region.

[29] Note that $\sum_L {}^{JL}r_i = 1$, since the coefficients are proportions of total production.

[30] In these tests, the two models were implemented using data from the 1960 interregional table to estimate base year trade and technical coefficients. Final demands were then calculated from the corresponding 1963 intraregional tables, and the 'accuracy' of the two models determined by comparing actual and projected regional output levels by sector. Several measures of accuracy were employed, but in almost every case the 'fixed shares of total sales to each region' model performed relatively poorly.

[31] See section 2.3 for a definition of the terms $^{JL}a_{ij}$, $^J X_j$ and $^J Y_i$.

[32] In order to avoid unnecessarily complicating the notation, this presentation assumes that the system of U regions does not trade with the 'outside world'. However, this is not necessary for the model's operation. See Leontief and Strout (1963).

[33] As will be discussed below, some of the approaches to the estimation of $^{JL}q_i$ suggested by Leontief and Strout (1963) can alleviate the magnitude of this problem. They do this by permitting the analyst to set $^{LJ}q_i$ equal to zero whenever she knows or suspects that $^{JL}X_i$ equals zero.

^{34}In an ideal system, where all regions are points, all commodity classes are homogeneous, and all shipments result from 'rational' decisions, cross-hauling would not occur. However, in reality none of the above conditions are actually met and any reasonable interregional trade flow model should be able to incorporate crosshauling (both actual and fictitious).

^{35}The terms $^{JJ}X_i$ and $^{LL}X_i$ appear on the RHS of equations (2.5.26) and (2.5.27), respectively, since (2.5.24) pertains to interregional flows only. Thus the subsidiary conditions $^{JJ}q_i = 0$ and $^{LL}q_i = 0$ reduce to zero the terms (^JD_i $^{JJ}q_i$) and (^LX_i $^{LL}q_i$), respectively.

^{36}We denote a base year magnitude by a bar, and a deviation from a base year magnitude by a delta. Also, new contants s and t are introduced to simplify notation:

$$^{JK}s_i = \begin{cases} ^JX_i(1 - {}^{JK}q_i) & \text{if } J \neq K \\ \\ ^JX_i - X_i + \sum_{K=A}^{U}(X_i {}^{KJ}q_i) \mid {}^{JJ}q_i = 0 & \text{if } J = K \end{cases}$$

$$^{KJ}t_i = \begin{cases} ^JD_i(1 - {}^{KJ}q_i) & \text{if } J \neq K \\ \\ ^JD_i - D_i + \sum_{K=A}^{U}(D_i {}^{KJ}q_i) \mid {}^{JJ}q_i = 0 & \text{if } J = K \end{cases}$$

The first order approximation of equation (2.5.29) can then be written as:

$$\sum_{L=A}^{U}({}^LD_i {}^{JL}s_i) - \sum_{L=A}^{U}({}^LX_i {}^{LJ}t_i) = 0 \qquad (i = 1, ..., n; J=A, ..., U; J \neq K)$$

where the n equations for region K have been dropped since they can be considered redundant. See Leontief and Strout (1963) and Polenske (1972) for further discussion.

^{37}Due to space and time limitations, these variants are not discussed here. The interested reader is referred to Wilson (1970), Gordon (1977) and Smith (1983, pp. 588-595).

[38] For all except the interregional trade flows, the data was assembled for 86 sectors (78 industries and 8 primary inputs) and 51 regions. However, the transportation data was available for only 44 regions. Further, trade flows were estimated for only 61 sectors, because the service sectors were assumed to serve the local market only.

[39] For further details on the nature of the MRIO mmodel, see Polenske et al (1972), Rodgers (1972), Scheppach (1972), Polenske et al (1974), Rodgers (1973), and Polenske (1980a).

CHAPTER 3
A COMPARATIVE COST AND INDUSTRIAL COMPLEX MODULE AS A COMPONENT OF AN INTEGRATED MULTIREGIONAL MODEL

3.1 Introduction

This chapter is concerned with the development of a comparative cost-industrial complex module which is capable of forming a component of an integrated multiregional model. In Section 3.2 an outline is given of the steps involved in conducting both comparative cost and industrial complex analyses for a system of regions. Section 3.3 provides a discussion of how the comparative cost and industrial complex approach can be used to provide projections of the regional production/trade patterns in cost-sensitive sectors while retaining the use of an interregional input-output model for projecting production/trade patterns for the other sectors of the multiregional economy. Finally, some concluding remarks are made in Section 3.4.

3.2 Outline of Comparative Cost and Industrial Complex Analysis
3.2.1 Comparative Cost Approach[1]

The comparative cost approach employs the framework of modern Weberian location theory (Weber, 1929 and Isard, 1956) to answer such questions as:

a) whether a given industry (with an established or anticipated pattern of markets for its output) can be expected to develop in a given region (with its particular set of resource and factor endowments); and

b) which regional distribution of production can be expected to exist or develop in a given industry (given its particular set of input requirements, and an established or anticipated pattern of markets for its output).

Regardless of which question is addressed, the general approach remains the same. The analyst constructs a systematic listing of cost differentials affecting the location pattern of the given industry, which then enables him to determine which region (or set of regions) could achieve the lowest total cost of assembling raw materials, converting these into finished product(s) and delivering this product to the market(s) under investigation. However, since the specific steps to be employed depend on whether the focus is on a single region (question (a)) or a set of regions (question (b)), each of these cases will be taken up in turn.

a. Single Region Applications

In determining whether a given industry (say industry j) can reasonably be expected to develop and prosper in a given region (say region J) the following steps are involved. See Figure 3.1.[2]

Step 1

From consultation with engineers and others familiar with the operations of the given industry, an efficient scale of operations is determined and information obtained on the inputs required in order to produce a unit of finished product at this scale. These requirements are expressed in physical units, such as man-hours of labor by type, kilowatt hours of electric power required to operate machinery, and tons of raw material by type.

Step 2

Identification is then made of those components of the given industry's total production and transport costs which currently exhibit significant variation between regions, and/or which can be expected to

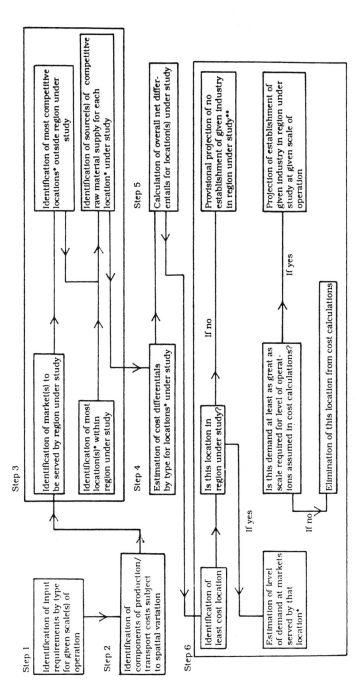

Figure 3.1 A Schematic Representation of the Comparative Cost Approach in Single Region Applications

- As discussed in the text, the same physical location may appear a number of times here depending on scale of operation, mix of inputs employed, sources of supply tapped, market area served, etc.
- This projection is provisional only, since it ignores the possibility that (due to agglomeration economies of various types) the given industry may be able to prosper as one component of an industrial complex

exhibit such variation over the relevant projection period. For example, such components may comprise transport costs on finished product(s), transport costs on raw material(s), the cost of extracting/processing raw material(s) at source(s), labor costs by type of employee, power costs by type of use, or taxation related costs or subsidies.

It may appear at first glance that a thorough comparative cost study should consider all components of cost in each region being compared. However, what is important in determining whether region J can support the establishment of a given industry within its boundaries, is not its total cost of operation there, but rather the differences between costs in region J and costs elsewhere. Thus, the analyst can ignore those components which do not vary significantly between regions. Not only does this lead to considerable saving of research time, but also makes the task of data collection easier.[2]

Step 3

Next the locations (regions) to be compared in subsequent steps must be selected. This involves:

a) identification of the market which could potentially be served by the given industry if it were to be established in region J;

b) identification of the location within J which is most likely to be competitive with other locations (regions) in serving the selected market;

c) identification of the other locations which are most likely to be competitive with J in serving the selected market. These should include both locations currently engaged in the given industry's production and those considered eligible to commence such production during the relevant projection period because of the significant cost advantage they possess. Say the actual locations to be considered are in regions K, L and M, and the potential locations in regions P, Q and R;[4] and

d) identification, for each location of the lowest cost source(s) from which the required inputs can be purchased.[5]

Step 4

The relevant cost differentials can now be estimated by type for the locations under study. This step is facilitated by recording the relevant data in a table such as Table 3.1. The locations (regions) selected for detailed comparison in Step 3 are listed along the lefthand tab. That is, starting from the top, we have the region under study (namely J), the regions with existing establishments in industry j (namely K, L, M,...), and the regions with potential for establishments in industry j (namely P, Q, R, ...). The industry j cost components, identified in Step 2 as subject to significant regional variation, are listed along the top of the table. That is, starting from the lefthand tab, we have transport related costs, raw material extraction/processing related costs, factor input and other costs. Taking each column in turn, the relevant cost estimates are filled in for each location. For example:

a) for the column relating to transport costs on finished product this involves procuring information on the costs of transporting one unit of finished product from each location to the given market (identified in Step 3);

b) for a column relating to transport costs on a given raw material (say, coal) this involves (1) procuring information for each location (region) on the cost of transporting one unit of coal from its given source (identified in Step 3) to that location, and (2) multiplying this estimate by the number of units of coal (identified in Step 1) required to produce one unit of finished product; and

Location	Selected Cost Components					Total of Selected Costs
	Transport Costs		Extraction/ Processing Costs	Factor Costs	Other Costs	
	Finished Product	Raw Materials 1....m	Raw Materials 1....m	Labour, Power, ...	Taxes....	
Region under Study J						
Regions with Existing Plants K L M . .						
Regions with Potential Plants P Q R . .						

Table 3.1 Locationally Significant Costs per Unit Output of Industry 1 at Selected Locations

c) for a column relating to costs associated with a given factor of production (say labor), this involves (1) procuring information for each region on the wage rate per man hour in industry j, and (2) multiplying this estimate by the number of man hours (identified in Step 1) required to produce one unit of finished product.[6]

Step 5

Adding up the costs expected at any location, the relevant total costs figure can be obtained. This is done in the last column of Table 3.1. Recall, that these figures do not represent the total costs of operation -- rather they refer to just those costs (identified in Step 2) which vary significantly among locations. From these figures, we calculate the net cost differential for region J relative to any other location (region).[7]

Step 6

For a given market (identified in Step 3) and a given scale of operations (identified in Step 1), it is obvious that the optimal location (region) will be the one which has the lowest total in the final column of Table 3.1. If this optimal location is not in region J then the analyst can conclude that, given his set of simplified assumptions, industry j cannot reasonably be expected to develop in region J.[8] If this optimal location is in region J, then the analyst must check that the level of demand for industry j's output in the given market is at least as great as the level of output required for efficient operation at the scale assumed for this location when conducting cost comparisons. If this condition can be met, then it can be concluded that if industry j was introduced into region J at the assumed level of operations, then it could be expected to prosper.

This completes the outline of the basic elements of the approach. However, modifications can be made to improve the reliability of its projections. For example:

a) Step 1 involved the specification of the input requirements associated with only one scale of operation; however the approach can be modified to consider a number of different scales of operation at each location. Specifically, a different set of input requirements can be specified for each scale of operation and the same physical location treated as a different location when it has a different scale plant associated with it. For example, "region J" may be replaced by "region J: small scale plant," "region J: medium scale plant," and "region J: large scale plant". Steps 2-5 can then proceed as outlined above. In Step 6, it must be recognized that if the given market cannot absorb the output associated with say a large scale plant in region J, then it cannot automatically be concluded that industry j will not be developed there. Rather, after "region J: large scale plant" is rejected as a location we need to identify the location which has the next lowest total in the final column of Table 3.1. If this location is also in region J, say "region J: medium scale plant" and the given market can support this scale of operation in J, then industry j could still be developed there. See Figure 3.1;

b) Step 1 involved the specification of only one set of input requirements for any given scale of operation; however different mixes of inputs can often be used depending on the type of technology employed. Further, because of such factors as inertia and differences in factor endowments it is reasonable to expect regional differences in input mixes for some industries. Where this is the case, the approach can be modified accordingly. Specifically, a different set of input requirements can be specified for each type of technology and the same physical location treated as a different location when it has a different technology associated with it. For example, "region J" may be replaced by "region J: technology type α", and "region J: technology type β". Steps 2-6 can then proceed as outlined above;

c) Step 3 involved the identification of only one market to be served by industry j in region J; however a number of markets may be accessible from J, and each market may have a different revenue potential and be associated with a different set of competitive locations. Where this is the case, the approach can be employed for each market separately;[9]

d) Step 3 involved the identification of only one location within region J; however, more could be considered. For example, "region J" may be replaced by "region J: location σ" and "region L: location δ". Steps 3-6 can then proceed as outlined above.

e) Step 3 involved the identification of only one source of raw material supply for each location -- namely the least cost source. However, the analyst may not be able to identify 'a priori' which source involves the least cost. For example, region J may be able to obtain a given raw material (say coal) from two sources. One source π is located close to J but is characterized by relatively high costs of extraction; the other source θ is located further away from J (and hence its use would incur greater transportation costs on coal) but can be extracted relatively cheaply. In order to determine which source should be tapped by industry i if established in region J, the analyst can conduct the cost comparisons in Steps 4-6 above for both "region J: source of coal at π" and "region J: source of coal at θ".[10,11]

This approach has been applied to the study of:

a) the iron and steel and petrochemical industries (see Isard and Cumberland, 1950[12] and Isard and Schooler, 1955[13] respectively) whose location patterns were found to be heavily influenced by transportation costs, provided the market served was large enough to support the optimum sized plant;

b) the aluminum industry (see Isard and Whitney, 1952)[14] which was attracted to regions offering a power cost advantage, because of its large power requirements per unit output; and

c) the textile industry (see Airov 1956, 1959) which was attracted to regions offering a labor cost advantage, because of its large labor requirements per unit output.

However, as suggested by Isard and Anselin (1982) it can also be applied to other sectors (such as pulp and paper, food and food products, and stone, clay and glass) whose location patterns can be explained by Weberian-type location principles.

b. Multiregion Applications

In projecting the future regional distribution of net production capacity expansions, and hence output, in a given industry (say industry j) the following steps are involved. See Figure 3.2.[15]

Step 1

The first step involves the projection of the expected net expansion K_j in production capacity in industy j for the nation as a whole. Generally, this involves:

a) the use of exogenous estimates[16] of future levels of national demand for industry j's output (namely $D_j(t)$), total foreign exports of industry j's output (namely $E_j(t)$), total foreign imports of industry j's output (namely $M_j(t)$), to derive an estimate (namely $D'_j(t)$) of future demand for industy i's output which needs to be satisfied from domestic sources. That is:

COMPARATIVE COST AND INDUSTRIAL COMPLEX 55

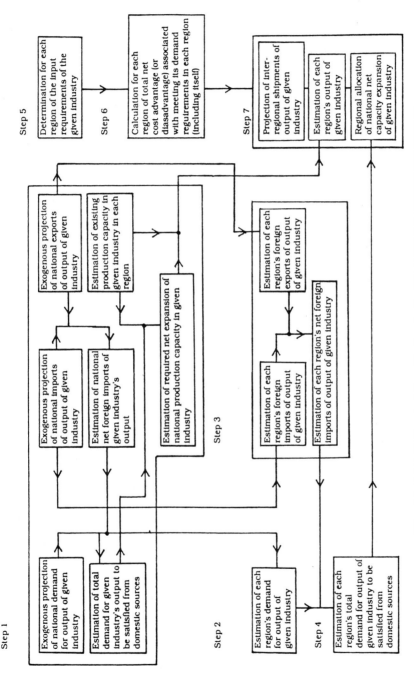

Figure 3.2 Schematic Representation of the Comparative Cost Approach in Multiregional Applications

$$D'_j(t) = D_j(t) - M_j(t) + E_j(t) \tag{3.2.1}$$

or $D'_j(t) = D_j(t) - M'_j(t)$

where $M'_j(t)$ = future level of total net foreign imports of output of industry i; and t = time period for which projections are being made.

b) the identification of the existing production capacity K_j in industry j within each region J (J=A,...,U) of the nation. An estimate can then be derived for the required net expansion in national production capacity K'_j. That is:

$$K'_j(t) = D_j(t) - \sum_{J=A}^{U} {}^J K_j \tag{3.2.2}$$

The subsequent steps then address the problem of how to allocate $K'_j(t)$ among the U regions.

Step 2

The second step involves the projection for each region J, of the future level of demand for industry j's output. Generally, this involves the allocation of the national estimate D_j among the U regions on the basis of the most reasonable allocator (or set of allocators) available. That is, for each region J we have:[17]

$$^J D_j(t) = {}^J \theta_j(t) \, D_j(t) \quad (J=A, ..., U) \tag{3.2.3}$$

where ${}^J D_j(t)$ = region J's demand for output of industry j; and

$^J\theta_j(t)$ = proportion of national demand for j allocated to region J.

Step 3

The third step involves the projection, for each region J, of the future level M'_j of net foreign imports of industry j's output. Generally, this involves

(i) the allocation of the national estimates of foreign exports (E_j) and imports (M_j) among the U regions on the basis of the most reasonable allocators available. That is:

$$^JM_j(t) = {^J\delta_j(t)} M_j(t) \quad (J=A, ..., U) \tag{3.2.4}$$

$$^JE_j(t) = {^J\Omega_j(t)} E_j(t) \quad (J=A, ..., U) \tag{3.2.5}$$

where $^J\delta_j(t)$ = proportion of national imports of industry j's output allocated to region J; and

$^J\Omega_j(t)$ = proportion of national exports of j allocated to J.[18]

and (ii) the subtraction of the export levels estimated in equation (3.2.5) from the import levels estimated in equation (3.2.4), to yield:

$$^JM'_j(t) = {^JM_j(t)} - {^JE_j(t)} \quad (J=A, ..., U) \tag{3.2.6}$$

Step 4

The fourth step involves the subtraction of the net import levels estimated in equation (3.2.6) from the total demand levels estimated in equation (3.2.3). This yields an estimate, for each region of the future demand $^J D'_j(t)$ for the output of industry j which needs to be satisfied from domestic sources. That is:

$$^J D'_j(t) = {}^J D_j(t) - {}^J M_j(t) \quad (J=A, ..., U) \tag{3.2.7}$$

Step 5

The fifth step involves the determination of the level $^J X_{ij}$ of each input i (i=1, ..., n) required per unit output of the given industry j at time t. In doing so, the analyst needs to take into account the potential impact of resource scarcities and technology changes. Also, where different regions can be expected to employ different technologies because of factor endowments, inertia, etc. then a different set of input requirements need to be estimated for each region J. These can then be summarized in a (U x n) matrix with typical element $^J X_{ij}$.

Step 6

The sixth step involves the calculation of the various cost differentials affecting the regional distribution of industry j. Where k cost components have been identified as subject to significant regional variation, this involves the construction of a set of U tables (each resembling Table 3.1) of dimension U x k. Each of these tables takes a particular region (say L) to be the market to be served, and examines for each region J (J=A, ..., U) the relevant cost advantages/disadvantages $^{JL}\rho_k$ (k = 1, ..., k) associated with delivering a unit of industry j's output to region L. Summing across the k columns of this table yields a U x 1 vector, with typical element $^{JL}\rho_k$ stating the total cost advantage (or disadvantage) enjoyed by producers in J when delivering a unit of output j to consumers in region L.[19] The region(s) with the highest total cost advantage(s) in serving region L can be easily identified by examining the elements of this vector.

Step 7

The seventh step involves:

a) the projection of $^{JL}X_j$ (J, L = A, ..., U), the interregional shipments of industry j's output which are required in order to ensure that available supply in each region matches the demand requirements estimated in Step 4. That is, it must be that:

$$\sum_{J=A}^{U} {}^{JL}X_j(t) = {}^{J}D'_j(t) \quad (L=A,...,U)$$

(3.2.8)

This matching process is assumed to take place in a cost efficient manner in the sense that each region L purchases its demand requirements from those regions identified in Step 6 as having the highest total cost advantage(s) in serving it. However, it is recognized that a certain level of cross-hauling is inevitable, so that the resulting pattern of shipments is not necessarily the one which minimizes total costs within the system as a whole.

(b) the estimation of each region J's output of industry j as:

$$^{J}X_j(t) = \sum_{L=A}^{U} {}^{JL}X(t) \quad (J=A,...,U)$$

(3.2.9)

(c) the estimation of $^JK'_j$ (J=A, ..., U), the resulting allocation of national net production capacity expansion K'_j among regions, as:

$$^JK'_j(t) = {}^JX_j(t) - {}^JK_j(t) \quad (J=A,...,U) \tag{3.2.10}$$

While this completes the outline of the basic elements of the approach when it is applied to many regions simultaneously, it should be clear that the approach can be modified in a number of places in a manner parallel to the discussion of pages 52-53.

This variant of the approach has been applied by Isard and Parcels (1977, 1977a and 1977b) to the study of the iron and steel, petrochemical and aluminum industries, respectively. In each of these applications, projections were made for the nine United States census regions for the years 1985 and 2000.

c. Evaluation of the Approach

In the above manner, an analyst could pursue comparative cost analysis for each industry in the region (or set of regions) under study. However, as the operational counterpart of Weberian location theory, the comparative cost approach is most useful for analyzing those industries (primarily manufacturing) whose location patterns approximate the set of conditions postulated by that theory. Perhaps most importantly, an individual producer in the industry should be able to be conceived of as: (a) locating at a point in space, (b) serving a market concentrated at a well defined point (or set of points), and (c) purchasing major inputs and/or factors of production also concentrated at a set of points.

Since many industries do not satisfy these conditions, comparative cost analysis needs to be supplemented by other approaches (for example, the economic base or input-output techniques discussed in Chapter 2) to yield output projections for these industries before it can be effectively used in a study of the complete industrial structure of a region (or set of regions). Further, even when applied to the study of industries for which it is ideally suited, the comparative cost approach has a number of limitations which need to be recognized when employing it for projection purposes.

One important set of limitations derive from the 'partial equilibrium' character of the approach. The approach can be claimed to be partial in two main respects:

a) Firstly, it concentrates on one particular industry and seeks to determine its 'optimal' location pattern while holding the location pattern of all other industries constant. However, it is clear that the location patterns of many industries are highly interdependent.[20] Where such interdependencies are considered too important to be safely ignored, the analyst may seek to adapt the approach in such a way that it can generate joint assessments of the feasibility of establishment in a given region[21] and/or joint projection of the regional distribution of expected net capacity expansions. However, another technique (namely the industrial complex approach to be discussed below) has been developed to handle such locational interdependencies in a more computationally efficient manner.

b) Secondly, it proceeds on the basis of an established or anticipated pattern of regional markets (demands), a given regional distribution of available raw materials and other input requirements, and a given set of ruling prices for inputs purchased and outputs sold in each region. However, where the postulated expansions in the given industry are large enough they could have a marked influence on either income, demand levels, output prices, or input costs in one or more of the regions under study. Where this is the case, the approach would need to be applied in an iterative fashion.[22] However, except in the simplest of situations, the resulting technique could quickly develop into a computational nightmare. It seems better to supplement the approach with other techniques specifically designed to analyze such interdependencies.

Another important set of limitations, which applies to many models but can bear repetition here, derive from the purely 'economic' character of the approach. Non-economic factors can often exert a strong influence on location decisions, and cause even the most carefully conducted comparative cost study to produce inaccurate projections. For example, when projecting the regional distribution of a given industry, the analyst needs to consider whether resistance[23] has developed (or can be expected to develop) to the establishment (or expansion) of their industry in one or more regions. When such resistance has been identified, the question needs to be asked of how strong is it, and what effect is it likely to have on location decisions. Reliable answers to these kinds of questions are not possible. However, the analyst may need to subjectively adjust his projections to reflect such resistances if they are to provide a useful guide to policy makers.[24]

Despite these limitations, the comparative cost approach has considerable potential in the Australian context -- particularly for use in projecting the level of foreign export demand for mineral and mineral-based products. See Appendix 3A for further relevant discussion.

3.2.2 Industrial Complex Approach[25]

Industrial complex analysis is concerned with the regional patterns of incidence (and growth) of groups of industrial activities which because of technical, production and marketing linkages generate significant economies to each activity when spatially juxtaposed. As discussed earlier, such interrelationships between activities suggest the need for joint assessments of the feasibility of establishment in a given region and/or joint projection of the regional distribution of expected national net capacity expansions.

a. Single Region Applications

In order to project which types of industrial complexes can reasonably be expected to develop and prosper in a given region, the following steps are involved. See Figure 3.3.

Step 1

First, we examine the resource (natural or manmade) and factor endowments possessed by the region, or lying in close proximity to it; and identify which of these could provide the region with a competitive advantage for industries seeking to establish there.

Step 2

Next, we identify the types of output which could potentially be produced from this set of resources. Any given resource may be able to be employed in a number of different production processes, and hence may be able to produce a large number of different outputs. Further, many of these outputs may be able to be used as intermediate inputs to other production processes -- yielding a still larger set of potential outputs. In short, there may be a large number of possibilities for utilization of each resource and it is often helpful to construct a flow chart which reveals the different input-output possibilities in terms of sequences or chains of products.[26]

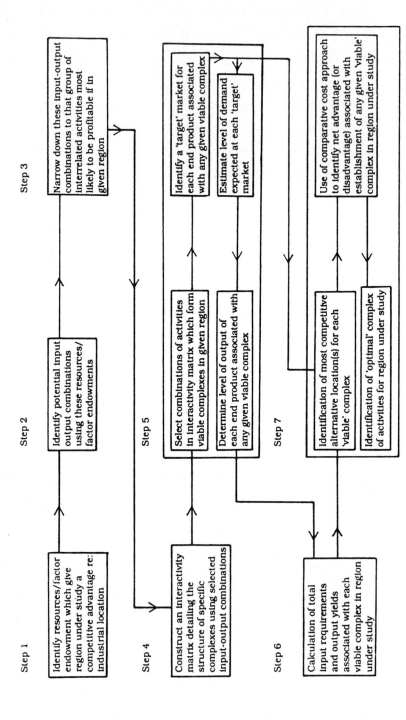

Figure 3.3 A Schematic Representation of Industrial Complex Approach in Single Region Applications

Step 3

The next step involves the use of technological, economic and other considerations to narrow down the set of potential input/output combinations to that 'complex of activities' most likely to be profitable if established in the given region. For example, some products may be eliminated since their production would require other inputs not available locally and very costly to import from other regions. Other products may be eliminated since even the minimum economic size plant would produce a scale of output in excess of what could be absorbed in local markets, and transport costs would be such that markets outside the region could not be penetrated. On the other hand, examination of trends in regional import/export data may reveal some products which should definitely be considered in subsequent calculations. For example, one product may currently be exported on a fairly limited scale, involve relatively minor transportation costs, and face a growing demand over the projection period. Other products may currently be imported on a fairly large scale, face a growing local demand, and yet involve relatively large transportation costs. And so on.

Step 4

The fourth step involves the determination of the detailed structure of specific complexes which could be formed on the basis of the selected set of input-output combinations. This is facilitated by construction of an (m x n) interactivity matrix A of the form depicted in Table 3.2.[27] At the head of each column are the individual or combined production processes (i.e., activities) that may be encountered in a specific selected complex.[28] As will be discussed below, for some processes (e.g., β and θ) a number of alternative prototypes are included here. The rows of the table refer to commodities which comprise either the outputs or required inputs of the various activities. Within the cells of the matrix, the inputs and outputs are expressed in physical units rather than in the monetary units typically employed in input-output analyses. When a commodity is used as an input to a particular process, this is indicated by a minus sign in front of the relevant entry; when it is yielded as an output then this is indicated by a plus sign in front of the relevant entry. For example, column 1 reveals the inputs and outputs associated with operating at a 'unit level' a hypothetical process (or activity) α.[29] That is, the required inputs include a_{11} units of commodity 1, a_{21} units of commodity 2, and a_{31} units of commodity 3; while the outputs yielded include a_{m1} units of commodity m, $a_{(m-1)1}$ units of commodity (m-1) and $a_{(m-2)1}$ units of commodity (m-2). One can also explore interrelations between the various columns, which permits the end products of any given activity to be traced back via a chain of intermediate products to their ultimate origins as a set of raw material and factor inputs.

However, there are other differences between the interactivity matrix depicted in Table 3.2 and a typical input-output matrix. For example:

a) those inputs/outputs which do not vary in direct proportion to the scale of operation of a given activity are excluded from the matrix. As will be discussed below these inputs/outputs are not ignored completely, rather they enter into the analysis at a later stage -- after the scale of operation of the different activities has been determined;

b) many of the cells within the matrix will be left blank, reflecting the fact that the commodity to which the row refers is not associated (either as a direct input or as a direct output) with the activity to which the column refers;

Activities Inputs/Outputs	Process α	Process β		...	Process θ		Process L
		Prototype 1	Prototype 2		Prototype 1	Prototype 2	
Commodity 1	$-a_{11}$						
Commodity 2	$-a_{21}$						
Commodity 3	$-a_{31}$						
.							
.							
Commodity (m-2)	$a_{(m-2)1}$						
Commodity (m-1)	$a_{(m-1)1}$						
Commodity m	a_{m1}						

Table 3.2 A Hypothetical Interactivity Matrix for Selected Input-Output Combinations

c) any activity (process) may be associated with multiple rather than single outputs; and

d) alternative processes are considered for performing the same functions within the complex. For example, there may be a number of different ways for converting commodity 1 (a raw material or factor input) into commodity 2 (an intermediate or finished product) -- each way requiring a different set of complementary inputs and yielding a different set of joint outputs. It is for this reason that some processes have more than one column, with each column representing an alternative prototype which could be employed.

Step 5

In the next step, technological, marketing and other economic considerations are employed to:

a) select from among the possible activities listed in the columns of Table 3.2 those which, in combination, might form a 'viable' complex when located in the given region;[30]

b) specify for each of the end products associated with any given 'viable' complex a target market which could effectively be served from the given region;

c) estimate the level of demand which can reasonably be expected at each of these target markets; and

d) determine the level of output of each end product which might be associated with any given 'viable' complex.

These four tasks must be performed simultaneously in the sense that the activities which might comprise a 'viable' complex depend on their respective scales of operation, and hence on the level of sales which the region might expect to capture at the target markets for the outputs of these activities. For example, if it is felt that the given region cannot effectively compete in outside markets for a given commodity, then a 'viable' complex might need to be put together in such a way as to restrict the output of this commodity to the level needed by other activities within the complex itself and/or which could be sold to other users (intermediate and final) in the given region.[31]

Step 6

The sixth step involves the calculation of the total inputs and outputs associated with each of the 'viable' complexes identified in Step 5. This calculation proceeds as follows:

a) For each activity i (i = 1, ..., n) in any given 'viable' complex, say complex I, a (1 x n) vector $^I X$ is constructed with typical element $^I x_i$ stating the multiple of the 'unit level' at which activity i must be operated in order to meet the requirements of complex I.[32]

b) Then, for commodities (h = 1, ..., m) involved in linear input-output relations we multiply the entries a_{hi}(+ or -) in the column corresponding to activity i in the interactivity matrix A by the multiple $^I x_i$ of its 'unit level' to obtain:

$$^I Z_i = A_i \, ^I X_i \qquad (3.3.1)$$

where A_i = a (m x 1) vector with typical element a_{hi} (+ or -); and

$^I Z$ = a (m x 1) vector with typical element $^I Z_{hi}$ stating for complex I the total input of commodity h to activity i if negative, and the total output of commodity h produced by activity i if positive.

The inputs and outputs of each commodity h (h = 1, ..., m) are then summed over all n activities in I to obtain:

$$^IZ_h = \sum_{i=1}^{n} {}^IZ_{hi} \quad (h=1,\ldots,m)$$

(3.3.2)

where ^IZ_h = the total input of commodity h to complex I if negative, and the total output of commodity h produced by complex I if positive.

These results can then be recorded in the first m rows of the first column of a table such as Table 3.3. The same set of calculations can be performed for other 'viable' complexes, and the first m rows of the remaining columns filled in.[33]

c) For commodities (h = m+1,...,q) involved in non-linear input-output relations we proceed as follows for any given 'viable' complex. Again complex I is used for illustrative purposes, and for each activity i (i=1,...,n) in I we have a set of equations of the form:

$$^IZ'_{hi} = ({}^IK_i/{}^*K_i)^{\beta_{hi}} * Z'^*_{hi}$$

(3.3.3)

$(0 \leq \beta_{hi} \leq 1;\ h = m+1,\ldots,q;\ i=1,\ldots,n)$

where ${}^*K_i$ = the capacity of activity (process) i when operated at the 'unit level';

${}^IK_i$ = the capacity of i when operated at the multiple ^IX_i of the 'unit level';

${}^*Z_{hi}$ = the total requirements for commodity h as an input to activity (process) i when operated at the 'unit level';

$^IZ_{hi}$ = the total requirements for commodity h as an input to i in 'viable' complex I; and

β_{hi} = an engineering-type factor which indicates how the requirements for commodity h vary with the scale of operation of

COMPARATIVE COST AND INDUSTRIAL COMPLEX

	Complexes / Inputs/Outputs	Complex I	Complex II	...	Complex X
Linear I-O Relations	Commodity 1 Commodity 2 . . . Commodity (m-1) Commodity m				
Non-Linear I-O Relations	Commodity (m+1) . . . Commodity q				

Table 3.3 Total Requirements and Yields of Selected Industrial Complexes

activity i over some meaningful capacity range. (It is assumed that $^*K_i$ and ^IK_i fall within this range).[34]

It is assumed that β_{hi}, $^*K_i$, and $^*Z_{hi}$ are known data.[35] Then with ^IK_i given, the requirements $^IZ_{hi}$ can easily be derived from (3.3.3). These requirements are then summed over all n activities in I to obtain the total requirements of complex I:

$$^IZ'_h = \sum_{i=1}^{n} {^IZ'_{hi}} \qquad (h=m+1,\ldots,q)$$

(3.3.4)

These results can then be recorded in the last q-m rows of the first column of Table 3.3 (with negative signs in front of them since they are inputs). This same set of calculations can be performed for other 'viable' complexes, and the last q-m rows of the remaining columns filled in.[36,37]

Step 7

With a table such as Table 3.3 constructed, the analyst is now in a position to proceed with the comparative cost framework outlined on pp. 45-54. The problem is to demonstrate, if possible, that one or more of the technically 'viable' complexes of activities selected for inclusion in that table could operate in the given region at a greater profit (lower costs) than elsewhere. In order to demonstrate this, the analyst must determine which alternative location(s) would be best able to serve the 'target' markets for the end products of these complexes. When doing this recognition of the following is required:

a) the most competitive alternative to locating a particular complex of activities in the given region may involve splitting these activities up among a number of different locations (regions),[38] and

b) the specific structure of the complex (factor proprotions, product mixes, process (activity) compositions, etc.) at the most competitive location(s) may differ from that considered most suitable for the given region.

With these location(s) identified, we next determine the net advantage (or disadvantage) associated with establishment in the region under study. The final results for any given complex of activities (say I) can be summarized as:

$$^I V = {^I \rho} \, {^I Z} + {^I \sigma} \, {^I Z'} \tag{3.3.5}$$

where $^I V$ = a scalar stating the net advantage (or net disadvantage) associated with establishing complex I in the region under study rather than at its most competitive alternative;

$^I \rho$ = a (1 x m) row vector with typical element $^I \rho_h$ stating the per unit net cost/revenue advantage (or disadvantage) associated with commodity h (h = 1, ..., m) when complex I is located in the region under study rather than at the most competitive alternative location(s);

$^I Z$ = a (m x 1) column vector of total input requirements and/or output yields whose typical element $^I Z_h$ was defined above;

$^I \sigma$ = a (1 x (q-m)) row vector with typical element $^I \sigma_h$ stating the per unit net cost advantage (or disadvantage) associated with commodity h (h = m+1,...,q) when complex I is located in the region under study rather than elsewhere; and

$^I Z'$ = a ((q-m) x 1) column vector of total input requirements whose typical element $^I Z_h$ was defined above.

The 'optimal' complex of activities for the region under study is then given as the one with the largest net advantage (or smallest net disadvantage). That is, as:

$$\max_{I} {}^{I}V \qquad (3.3.6)$$

Although many other refinements can be introduced into the study, the above discussion suffices to illustrate how the industrial complex approach can be used to determine the feasibility of establishing a given set of activities in a particular region. The approach has been employed, for example;

a) to examine the possible introduction of an oil refinery-petrochemicals-synthetic fiber complex into the Puerto Rico economy (Isard and Schooler, 1955 and Isard et al ,1959);

b) explore the feasibility of expansions to the non ferrous metallic mineral products based complex in Northern Ontario (Hodge and Wong, 1972); and

c) identify and analyze potential chemical manufacturing complexes (Sommerfeld et al, 1977).[39]

b. Multiregion Applications

The approach can be extended to multiregional applications; that is, for use in projecting the regional distribution of net capacity expansions in a set of interrelated industrial activities. This extension will not be presented in detail here. However, the general steps involved would be:

a) use of exogenous estimates of national demand for, and net foreign imports of, the end products of each of the activities comprising the complex to derive an estimate of national demand for each of these end products which must be satisfied from domestic sources;

b) estimation of existing production capacity in each of the given activities in each region, and the use of these estimates to derive the required net expansion of national production capacity in each activity;

c) allocation of the exogenous national demand and net foreign import estimates for each end product among regions using the most reasonable allocators available;

d) use of the estimates derived in (c) to calculate for each region the demand for each end product which must be satisfied from domestic sources;[40]

e) identification of those resources and/or factor endowments in each region which provide it with a competitive advantage (or disadvantage) for the establishment/expansion of each activity within its boundary;

f) completion of Steps 3-7 outlined on pp. 65-73 above for each region separately, except that for each region J we now derive a net advantage (or net disadvantage) I^V with respect to each other region L in the system;[41]

g) use of a 'cost efficiency' principle[42] in the projection of interregional shipments of commodities (intermediate as well as end products) which are required in order to ensure that the available supply of each activity's end products in each region matches its demand requirements estimated in (c);

h) use of the estimates derived in (g) to calculate each region's output of the end products of each activity; and

i) use of the output projections derived in (h) and the existing production capacity estimates derived in (b) to yield, for each activity comprising the complex, the required regional allocation of national net production capacity expansion.[43]

c. Evaluation of the Approach

In conclusion, the industrial complex approach can be a useful technique when applied to the projection of the interregional distribution of a set of industrial activity combinations, or to the identification and evaluation of profitable activity combinations for any given region. It is able to capture:

a) the effects of production, marketing and other interrelations between activities -- which a set of industry-by-industry comparative cost studies (without major modification) cannot; and

b) the effects of process (technology) substitution possibilities, nonlinear input-output relationships and multiple outputs for each activity comprising the complex -- which a strictly linear input-output model cannot.

However, there are limits to the usefulness of the industrial complex approach. For example, since it employs a comparative-cost approach to determine the relative attractiveness of each region for the establishment/expansion of the various activities comprising a complex, it too is most useful for analyzing those industrial activities (mainly manufacturing) whose location patterns approximate the set of conditions postulated by Weberian location theory. Even when applied to the study of sets of activities for which it is ideally suited, other limitations arise. Some are similar to those examined on pp. 60-62 in relation to the comparative cost approach, namely:

a) it concentrates on a particular cluster of industrial activities and seeks to determine their 'optimal' structure and location pattern while holding the structure and location pattern of all other industries constant. Since the location patterns of different clusters of activities are often highly interdependent, the analyst may seek to extend the approach to determine the 'optimal' structure and location pattern of "clusters of clusters of activities". However, such extensions will not be discussed here since (except in the simplest of situations) the resulting technique could quickly develop into a computational nightmare; and

b) it proceeds on the basis of a given pattern of markets (demands) for each of the end products produced by the complex, a given regional distribution of available raw materials and other input requirements for each activity comprising the complex, and a given set of prices for the various inputs purchased and outputs sold by the complex in each region. However, where the postulated expansions in one or more of the activities comprising the complex are large enough they could have a marked impact on the relevant demand levels, output prices, and input costs in one or more of the regions under study.[44]

Other limitations are different. For example, the approach is data hungry with the result that where:

a) the production and/or marketing interrelations among activities are relatively insignificant, or

b) the economies achieved by spatial juxtaposition are largely non-quantifiable; then an industry-by-industry comparative cost approach may comprise a more cost effective research tool -- especially when appropriate qualifying statements are provided together with the projected activity levels in each region.

But, as suggested by Isard et al (1960) the real merit of the industrial complex approach is not as a substitute for the techniques already discussed; but rather as a complement to these techniques. For example, the analyst may employ an industrial complex approach in projecting variations in the level of certain industries, which are highly interrelated (e.g., mineral processing and metal-based manufacturing activities)[45], while permitting variation in other industries (e.g. retailing and other service type industries not suited to industrial complex analysis) to be estimated via input-output or economic base techniques. Variation in still others may be handled via the optimization approach of the linear programming module to be discussed in Chapter 4.

3.3 Integration of Comparative Cost-Industrial Complex Analysis and Multiregional Input-Output Models

The linkage of comparative cost-industrial complex and interregional input-output approaches was first suggested by Isard (1960) and subsequently elaborated upon by Isard et al (1960).[46] A schematic representation of this integrated approach is given in Figure 3.4; where 'traditional' final demand comprises consumption, investment, and government expenditure as well as export revenue, while 'non-traditional' final demand is as defined in step 3 below. The steps involved are:

Step 1: *Identification of 'Reasonable' Projections of Relevant National Variables*

The first step involves the identification of exogenous projections of the following system (national) variables: output by sector, employment by sector, final demand by type by sector, and population by age/sex. These projections should be consistent with the analyst's basic assumptions about the future of the system; and could come from the operation of existing national (system) input-output, econometric and demographic models.

Step 2: *Derivation of Initial Traditional Final Demand Projections*

The next step involves:

COMPARATIVE COST AND INDUSTRIAL COMPLEX

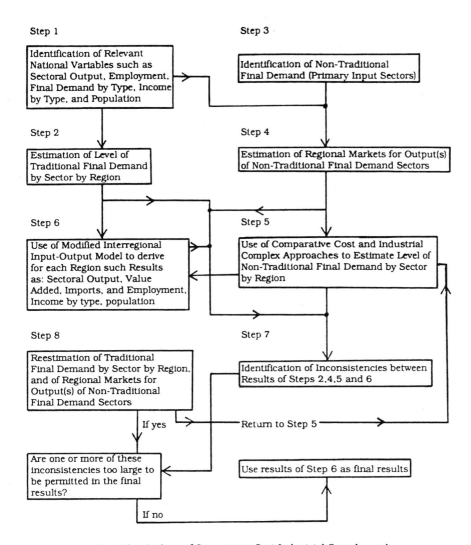

Figure 3.4 Linkage of Comparative Cost-Industrial Complex and Interregional Input-Output Models

(i) the disaggregation by region of the national (system) final demand estimates estimated in Step 1. This is done by using regional share coefficients calculated on the basis of current data (on such variables as regional population and income) via the use of trend analysis, relative growth charts, results of consumption and government expenditure analysis, and diverse coefficients and curves (such as location quotients, and localization curves);[47] and

(ii) for each region, say J, the resulting final demand estimates need to be adjusted by:

-subtracting the level of imports to J expected to be purchased from other regions L (L = A, ..., U) in the system; and

-adding the level of exports from J expected to be sold to other regions L in the system. The required levels of adjustments can be estimated via the use of the same data and/or procedures used to generate trade coefficients for the interregional input-output submodel. This yields a projection of the level of 'traditional' final demand by type by sector by region.[48]

Step 3: *Identification of Cost-Sensitive Industries*

The third step involves the identification of those industries (or parts of industries), say $i=1,...,n$, whose locational patterns are to be projected via the use of comparative cost or industrial complex analysis. Next, for each of these industries (or parts of industries):

(i) the column representing its input requirements by sector by region, and

(ii) the row representing its output sales by sector by region, are removed from the interindustry section of the traditional interregional input-output transactions table [49] and placed within a set of 'non-traditional' final demand columns and 'non-traditional' primary input rows, respectively.

Specifically, we now have a modified input-output table of the form depicted in Table 3.4.[50] In the section of this table designated block I, we have an interregional interindustry transactions matrix for those endogenous producing sectors ($i=1,...,n$) whose locational patterns are to be determined via the operation of the input-output submodel. In the remaining columns of the table the non-traditional ($i=1,...,n$) and traditional ($f=1,...f$) final demand sectors are listed. Their input requirements from the various industries (or parts of industries) remaining in the traditional interindustry transactions matrix are noted in the cells of blocks II and III, respectively; while their input requirements from those producing sectors ($i=1,...,n$) whose locational patterns are to be determined via the operation of the comparative cost-industrial complex submodel are noted in the cells of the blocks V and VI, respectively. In the section of the table designated block IV, we have the input requirements of the endogenous producing sectors from those producing sectors placed within the non-traditional final demand columns. Finally, in the remaining rows of the table the traditional primary input categories ($v=1,...,v$) are listed. Thus, in blocks VII, VIII and IX we have the traditional primary input 'requirements' by endogenous producing sectors ($i = 1, ..., n$), and producing sectors placed within the non-traditional final demand columns ($i = 1, .., n$) and the traditional final demand sectors ($f = 1, ..., f$), respectively.

Step 4: *Derivation of Initial Market Projections for Cost-Sensitive Industries*

The fourth step involves the estimation of the level of regional demands (intermediate and final) for the output(s) of each of the

Outputs \ Inputs	Intermediate Sectors		Final Demand Sectors				Total Gross Output
			Non-Traditional		Traditional		
	J	L	J	L	J	L	
	1,...,n	1,...,n	1,...,n	1,...,n	1,...,f	1,...,f	
Intermediate Sectors J 1...n L 1...n	Block I		Block II		Block III		
Non-Traditional Primary J 1...n L 1...n	Block IV		Block V		Block VI		
Input Sectors Traditional J 1...v L 1...v	Block VII		Block VIII		Block IX		
Total Gross Outlays							

Table 3.4 Modified Interregional Input-Output Transactions Table

industries (or parts of industries) identified in Step 3. As discussed on pp. 56-58, this can be done by:

(i) identifying the national demand (intermediate and final) for the output of the non-traditional sectors, after excluding those intermediate demands which derive from the non-traditional final demand sectors themselves;

(ii) adjusting these demand levels for net foreign imports; and

(iii) allocating the adjusted demand among regions on the basis of the best allocators available.[51]

Step 5: *Derivation of Initial Non-Traditional Final Demand Projections*

Given the pattern of markets identified in Step 4, the next step involves:

(i) the use of multiregional variants of the comparative cost and/or industrial complex approaches[52] to derive estimates for each region of the level of each industry (or part of industry) placed within the non-traditional final demand column(s); and

(ii) the use of the results of these comparative cost and industrial complex analyses to modify (where required) estimates of some of the input-output type coefficients in the non-traditional final demand columns and non-traditional primary input rows of Table 3.4.[53]

The cells within the 'non-traditional' final demand columns of the modified input-output matrix (namely blocks II, V and VIII in Table 3.4) can then be filled in.

Step 6: *Derivation of Initial Output Projections*

Now that the 'traditional' and 'non-traditional' final demands by sector by region have been estimated (in Steps 2 and 5, respectively) the interregional input-output submodel may be operated. Such operation yields for each region the corresponding output levels for those endogenous producing sectors (i = 1, ..., n) remaining in the structural matrix. Given these output levels, it then becomes possible to fill in the cells of blocks I, IV and VII of Table 3.4.[54]

Step 7: *Identification of Inconsistencies in Output Projections*

Generally speaking, the results derived from an initial run of Steps 1-6 will give rise to a number of inconsistencies. For example:[55]

(i) For each non-traditional final demand sector, the regional pattern of outputs derived at the end of Step 5 may be inconsistent with the set of regional markets initially estimated (or assumed) in Step 4;

(ii) For each region, the output levels for the non-traditional final demand sectors derived at the end of Step 5 may be inconsistent with the set of requirements for the outputs of these sectors yielded via use of the interregional input-output submodel in Step 6. (The total of these requirements is recorded for each sector in the appropriate row of the total gross output column at the extreme right of Table 3.4);[56]

(iii) another set of discrepancies arise for each region when we compare both:

-its pattern of non-traditional final demand sectors' outputs as derived at the end of either Step 5 or 6, and

-its pattern of endogenous producing sectors' outputs as derived at the end of Step 6, with the level of its traditional final demand by type by sector as estimated at the end of Step 2.

For example, for region J the level of <u>realized</u> household income recorded in the appropriate cell of the total gross output column of Table 3.4 (and generated as the sum of payments to households by endogenous producing sectors and traditional and non-traditional final demand sectors) may differ from the level of household income <u>anticipated</u> (explicitly or implicitly) for region J when calculating the level of its household consumption expenditure in Step 2. Similarly with respect to realized government receipts by region and the level of anticipated revenues upon which the government expenditure calculated in Step 2 was based, and so on.

Step 8: *Elimination of Major Inconsistencies*

If any of the inconsistencies identified in Step 7 are judged to be too large to be permitted in the final projections, then the analyst must re-estimate:

-the level of traditional final demand by sector by region, and/or

-the pattern of regional markets (demands) for the output of nontraditional final demand sectors.

The results derived at the end of Step 6 provide the basic information to be employed in this re-estimation process. For example:

(i) the regional pattern of expenditures for the household consumption final demand category can be re-estimated on the basis of payments to households (by endogenous producing sectors and traditional and non-traditional final demand sectors) consistent with the gross output levels recorded in the final column of Table 3.4.[57] Expenditures falling within other traditional final demand sectors can be re-estimated in a similar manner; and

(ii) the regional pattern of markets for the output of any given non-traditional final demand sector can be re-estimated by calculating the requirements for its output (by endogenous producing sectors, and traditional and non-traditional final demand sectors) consistent with the gross output levels recorded in the final column of Table 3.4.[58]

Once the revised sets of traditional final demands and markets for non-traditional final demand sectors' outputs are obtained, Steps 5-7 are repeated. That is new comparative cost and industrial complex analyses are conducted to establish new (more logical) levels for non-traditional final demand sectors in each region. The two sets of revised final demands are then employed in a rerun of the interregional input-output submodel, and a new set of results recorded in Table 3.4.

Such reruns may be required for a number of rounds until all major discrepancies are eliminated. However, no formal proof has yet been provided that this rerun procedure will lead to successively smaller and smaller discrepancies; and there is insufficient experience with operating such an integrated model for any empirical results to be reported here. Nevertheless, Isard et al (1960) argue that, by introducing a number of constraints reflecting resource scarcities and community attitudes in the different regions, the sophisticated analyst should be able to force the results of the successive runs to converge in a reasonable manner.[59]

3.4 Concluding Remarks

This chapter has dealt with the use of a comparative cost-industrial complex module as a component of an integrated multiregion model. In particular, linkage with an interregional input-output module has been discussed at length.

Although the operationality of this module has yet to be proven in the Australian context, we feel confident that such can be done. Given the growing reliance on minerals and mineral-based products as major foreign export earners, it is imperative that an attempt be made to project Australia's competitive standing in the major markets for these products. The inclusion of a comparative cost-industrial complex module within the integrated model offers a reasonable path towards achieving the required projections.

NOTES

[1] See Isard et al (1972), (1960), Isard et al (1972), and Isard (1975) for general outlines of this approach.

[2] The sequence of steps presented in this Figure should not be viewed as rigid, however all steps would be pursued at some stage during a study employing a comparative cost approach.

[3] For example, where a survey is conducted of selected firms in the given industry, it is often easier to get cooperation when only relatively few individual production cost items are requested.

[4] It may appear at first glance that a thorough study would consider all locations (regions) which could serve the given market. However, what is important is whether region J can be cost competitive with its 'least cost' rival. If it was known beforehand which location this was, then only two regions (locations) need be compared.

[5] For actual locations (K, L and M) these would normally be the existing sources of supply, however, we can allow for the possibility that other sources will be tapped during the relevant projection period.

[6] It should be noted that in considering an element of production cost which does vary regionally, it is often possible to calculate the amount of this difference between regions without knowing its absolute regional level. These estimates could be inserted in Table 3.1, and the analysis conducted in the same manner. See Isard et al (1960) for further discussion.

[7] It is at this point that the parallel between the comparative cost approach and modern Weberian location theory is most easily seen. Addition of the elements within the transport cost columns for each region would allow the total transport cost differentials to be calculated. (This resembles the process by which isodopanes are derived from isotims). Then the 'optimal' location on the basis of transport cost considerations alone can easily be identified as the one with the lowest transport cost differential. (This resembles the identification of the point at the center of the innermost isodopane as the point of minimum total transport costs). We can then consider each other column (locational factor) in turn. For example, we can identify the 'optimal' location on the basis of both transport cost and labor cost differentials and see if this is different from the transport cost based optimum. (This resembles the use of the concept of a 'critical isodopane' as the locus of points where the labor cost savings are just offset by increased transport costs. If no cheap labor cost location falls within its relevant critical isodopane, then the point of minimum transport costs remains the optimum location. If more than one cheap labor location falls within their respective critical isodopane, then the one which lies furthest inside its critical isodopane becomes the new

'optimum' location). We can also consider a number of additional columns (locational factors) simultaneously. For example, we can identify the 'optimal' location on the basis of transport, labor and power cost differentials and see if this is different from the transport cost based optimum. (Here the relevant critical isodopanes are defined as the locus of points where the net labor and power cost savings are just offset by transport cost increases). And so on.

[8] This conclusion should be recognized as provisional, since it ignores the possibility that (due to agglomeration economies of various types) industry i may be able to develop and prosper as one component of an industrial complex in region J. See the discussion in section 3.2.2.

[9] Where scale economies are an important factor determining location patterns in the given industry then the results of these separate studies may need to be evaluated jointly, since region J may need to be the least cost location in all markets to which it has access for the assumed scale of operations to be viable.

[10] Alternatively, the source which lies closest to region J may consist of low quality coal with the characteristic that more of it is required to produce a unit of finished product. When this is the case, different input requirement vectors are associated with region J, depending on which source is tapped. However, it is clear that the approach can be modified to take this into account also.

[11] Of course, it is possible to introduce a number of these complicating factors simultaneously. However, after a point the increased data requirements become prohibitive and the additional computations become cumbersome to perform.

[12] In this application, the objective was to determine whether the New England region could support an integrated iron and steel works capable of serving the Boston market. Transport cost differentials, on the major raw materials (iron ore and coal) and finished steel products were found to be the most important cost components subject to regional variation. Two locations within New England were selected for study, and two alternative sources of iron ore considered. Although the New England location was found to enjoy a net transportation advantage over its competitors, it was uncertain whether the demand for steel products in the Boston/New England market was sufficient to absorb the output of a minimum economic size integrated iron and steel plant. Therefore, the establishment of such a plant was not recommended.

[13] In this application, the cost comparisons were made between sites in the Gulf Coast region and a number of sites central to the market area (namely Cincinatti) selected. Since the output of the optimal sized plant would be absorbed by the part of the market area accessible to each site, transport costs were found to be the most important cost component subject to

regional variation. Because of its large natural gas reserves at that time, Gulf Coast locations minimized transport costs on raw materials; while the market locations (practically speaking) avoided transport costs on finished products. When the detailed cost comparisons were performed, a site in Munroe, Louisiana was found to be the most favorable location.

[14] In this application, the regions selected for cost comparison included New York City (a market location) and the Pacific Northwest (because of its cheap hydro-power resources). The transport and power cost components were the ones identified as most critical in determining location patterns. Although the New York City location was found to have a net transport cost advantage, this was overshadowed by the significant power cost advantage enjoyed by sites within the Pacific Northwest.

[15] Once again, the sequence of steps presented in this Figure should not be viewed as rigid. See Isard and Parcels (1977, 1977a, 1977b) for examples of multiregional studies employing a comparative cost approach.

[16] These estimates could be derived from a national econometric/input-output model, or perhaps from a set of integrated national econometric/input-output models. See Isard and Anselin (1982), and Isard and C. Smith (1982, 1982a).

[17] For example, for many industries future population levels $^J P(t)$ may be used as a crude allocator. In this case, $^J\phi(t) = {}^J P(t)/P(t)$. However, regardless of which allocator is used it must be that $\sum_J {}^J\phi(t) = 1$.

[18] Note that we require that $\sum_J {}^J\delta = 1$ and $\sum_J {}^J\Omega = 1$.

[19] Where it is obvious that not all regions can support a given industry (for example, because it requires use of a resource which is highly localized and non-transportable), then this step may be conducted for a restricted set of regions. See Isard and Parcels (1977b).

[20] For example, one industry may sell part of its output to (or purchase one of its inputs from) the next, and hence benefit from transport cost savings when located together at the same location. Alternatively, if two industries requiring the same input were to locate at the same location, then a third industry producing this input close by that location could increase its scale of operations and perhaps make it available at a lower cost because of economies of scale. Or perhaps all three industries located next to each other would employ sufficient workers to justify the establishment of a government-financed technical college nearby, which may then make the training of skilled and semi-skilled personnel less expensive for each industry. And so forth.

[21] That is, while industry j may not be able to develop and prosper in a particular region if it were to be the only industry introduced, if another industry (or set of industries) to which it was closely related were also introduced then the negative conclusion re: industry j's viability in the particular region may be able to be reversed.

[22] That is, an initial set of markets, prices, etc. would be assumed, the cost calculations performed and expected output sales/input purchases in each region derived. Any obvious inconsistencies between these results and the initial assumptions would then be identified, the assumptions reset where this is considered necessary, new results derived, and so on.

[23] Such resistance may be formal (in the sense that it has become institutionalized) or informal. Also, it may stem from other businessmen, social groups or household residents. See Isard et al (1960) for further discussion.

[24] See Isard and Parcels (1977b) for an attempt to conduct such adjustments for hydropower expansion in the Pacific region of the United States.

[25] See Isard et al (1959), Isard et al (1960), and Isard (1975) for general outlines of this approach.

[26] Such a flow chart for the possible intermediate and end products from oil and natural gas is given in Isard et al (1959, pp. 30-31), Isard et al (1960, pp. 380-381), and Isard (1975, pp. 440-441).

[27] An example of such a table for oil refinery, petrochemical and synthetic fiber activities is given in Isard et al (1959, pp. 40-49). A summary version of this table is also reproduced in Isard et al (1960, pp. 384-385) and Isard (1975, pp. 472-473).

[28] Processes (activities) which are technically infeasible or economically unreasonable for the given region are excluded from such a matrix. Further, since only those inputs/outputs which lead to significant spatial variations in costs or revenues need be considered in determining whether the given region can support a particular activity, such a matrix does not include all inputs/outputs associated with any given activity.

[29] The selection of the 'unit level' for any process (activity) is arbitrary. The investigator usually defines it at a scale so as to facilitate computation and/or understanding of the problem. See Isard et al (1959) for further discussion.

[30] Note that a number of alternative 'viable' complexes (say I, II, ..., X) may be identified here.

[31] However, in some cases an 'oversupply' of a given commodity cannot be avoided regardless of the mix of activities (processes) included in the complex.

[32] These requirements are a function of the output levels that were identified in Step 5 for each end product associated with that complex.

[33] Note that many of the cells in this portion of Table 3.3 will contain entries of zero, or numbers close to zero. Where this is the case either (a) no inputs or outputs of the given commodity are associated with the corresponding complex, or (b) the corresponding item is an interemediate commodity whose output from one or more activities matches its use in one or more other activities. See Isard et al (1960) for further discussion.

[34] The lower limit to this range represents the minimum scale of output below which the activity must generally be operated with different and less efficient methods; while the upper limit represents the maximum scale of output beyond which (because of sharply rising costs of operation) duplication of plant or process units is preferable to expansion of a single unit or plant capacity. See Isard et at (1959) for further discussion.

[35] For example, by defining the 'unit level' of operation of activity process i in such a way that it coincides with that employed in an existing plant $*K_i$ and $*Z_{hi}$ are easily derived. This may also facilitate the estimation of β_{hi}, since it is then defined for a range around the existing capacity $*K_i$.

[36] See Isard et al (1959, pp. 90-91), Isard et al (1960, pp. 390-391) for an example of such a table. In this study, capital and labor were the two "commodities" subject to nonlinear input-output relations.

[37] The above presentation assumes that if a given commodity h is involved in nonlinear input-output relations in one activity i in one 'viable' complex I, then it is involved in such nonlinear relations in all n activities comprising that complex -- and in all other 'viable' complexes being considered. However, this assumption is made to simplify the notation and need not be employed in any given application.

[38] For example, Isard et al (1959) compared a full oil refinery-petro-chemical-synthetic fiber complex in Puerto Rico with a combination of split complexes on the mainland of the United States. The mainland combination comprised: (a) oil refinery-petro-chemical activities in the Gulf Coast region; and (b) synthetic fiber activities in the textile producing regions of the South.

[39] Alternative approaches for examining industrial complexes have been developed (see Czamanski and Ablas (1979) for a review of these approaches and Czamanski and Czamanski (1976) and Roepke et al (1974)

for examples of their application). Generally speaking, these approaches analyze the structure of a given economy and use statistical techniques such as multivariate analysis and graph theory to identify industrial clusters (that is, groups of industries connected by flows of goods and services stronger than those linking them to other sectors of the economy) and/or industrial complexes (that is, industrial clusters whose component industries also have a significant similarity in locational patterns.) However, since the objectives of these studies are different from those of this book, they will not be reviewed here.

[40] Since the above four steps are parallel to those discussed on pp. 54-58 for the multiregion application of the comparative cost approach, the reader can refer back to there for more detailed discussion of the nature of these steps.

[41] For practical purposes, these comparisons need only be made with those regions which could be competitive with J in the production of a given end product.

[42] See Smith (1978, 1981) for a discussion of the rationale underlying this principle.

[43] Since the last three steps parallel the final step of the multiregional application of the comparative cost approach, the reader can refer back to there for more detailed discussion of the nature of these steps.

[44] As discussed on pp. 58-60, it is possible to modify the technique in such a way as to reduce the seriousness of this limitation. For example, rather than assuming a given pattern of resource availabilities at set prices the analyst may be able to develop a set of meaningful regional supply schedules. Similarly, she may have access to regional demand schedules for each finished product and hence be able to drop the assumption of a given pattern of markets at set prices. However, the technique can become impractical if these assumptions are dropped for a number of inputs/outputs simultaneously.

[45] See Appendix 3A for relevant discussion, since here we explore the use of the industrial complex approach for projecting the mix of extraction, benefication, refining, smelting and fabrication activities which can be successfully operated in Australia (recognizing the export-orientation of most of these activities).

[46] Here, the integrated model was designated Channel I. See also Isard and Anselin (1980).

[47] See Isard et al (1960, Chapter 12) for further discussion.

[48] Such results could, for example, be derived from a highly disaggregated multisectoral, multiregional econometric model of a bottom-up character. However, the data required for adequate implementation of such a model is generally not available.

[49] Where it is meaningful to treat only part of an industry by Weberian location analysis, the remaining part is left within the inter-industry section of this table. Its location pattern is then determined in accord with the spatial relations embodied in the interregional input-output submodel.

[50] Although this table assumes only two regions, namely J and L, the format of the table does not change when more regions are considered.

[51] These allocators could be simply based on readily available materials (such as regional population) or be derived from a more complex process involving the use of the trade coefficients derived from commodity flow studies or non-survey techniques.

[52] See pp. 54-60 and 73-74 for a discussion of the nature of these variants and the circumstances under which one approach should be used in preference to the other.

[53] Recall that Tables 3.1 and 3.3 refer to inputs and/or outputs which are likely to cause significant regional cost/revenue differentials. Thus, even when these input requirements are disaggregated by region of origin, they provide insights into only some of the cells within the corresponding non-traditional final demand column. Similarly, for the regional pattern of output sales.

[54] For example, the cells in block I are derived by multiplying these outputs by the relevant interregional input-output coefficients. In somewhat the same manner, the cells in blocks IV and VII can be derived.

[55] For further discussion of the nature of these discrepancies, see Isard et al (1960, pp. 582-600).

[56] As discussed on pp. 60-62 and pp. 75-76, these two sets of inconsistencies derive from the 'partial equilibrium' character of the comparative cost and industrial complex approaches when applied to a single industry (or complex of activities). This problem becomes even more severe when these approaches are applied to a number of industries (or complexes or activities) simultaneously -- for here the interdependencies between industries (or complexes or industries) placed in the non-traditional final demand sectors cannot be adequately captured.

[57] Those payments currently recorded in Table 3.4 are consistent with the gross outlays recorded in the final row of that table. It is because of inconsistencies between gross outlays and gross outputs (discussed in Step 7 above) that this re-estimation process will generally produce revised regional expenditure estimates. See Isard et al (1960, pp. 596-597) for further discussion.

[58] Those requirements currently recorded in Table 3.4 are consistent with the gross outlays recorded in the final row of that table. It is because of inconsistencies between gross outlays and gross outputs (discussed in Step 7 above) that this re-estimation process will generally produce revised regional market estimates. See Isard et al, (1960, pp. 596-597) for further discussion.

[59] For example, such constraints may be used to disallow results which imply divergences from base year magnitudes which can be considered excessive from either a technical or political standpoint.

[60] Note also that the comparative cost-industrial complex technique can effectively be linked to a balanced regional input-output model. For a discussion of the steps involved in this linkage see Isard (1960), Isard et al (1960), Isard et al (1969) and Smith (1983, Chapter 4).

APPENDIX 3A:

EXTENSION OF COMPARATIVE COST AND INDUSTRIAL COMPLEX ANALYSIS TO INTERNATIONAL APPLICATIONS

Overseas exports account for around 16 percent of Australia's gross domestic product, with some states more and others less 'export-oriented' than the nation. As a result, the export component of the input-output final demand vector requires a good deal of attention. One could use time trends to project each state's overseas exports by commodity. However, Austrialia has recently experienced what some term a "mining boom" with discoveries of major reserves of iron ore, coal, bauxite, mineral sands, nickel, uranium and so on. In fact, we have now become major world producers of most of these minerals, some of which were previously imported. Under these circumstances, the use of time trends for projecting exports of many minerals and mineral-based products would be highly misleading.

Since many of these minerals are transport cost sensitive, and their refining and smelting tend to be energy cost sensitive, the use of the multiregion version of the comparative cost-industrial complex approach developed in Chapter 3 is suggested. That is, for each of the major markets for Australian mineral products we compare the cost differentials associated with the use of Australian exports compared with the most competitive alternative source. These costs differentials then provide the basis for projecting Australian exports to these markets.

CHAPTER 4
A LINEAR PROGRAMMING MODULE AS A COMPONENT OF AN INTEGRATED MULTIREGIONAL MODEL

4.1 Introduction

This chapter is concerned with the development of a linear programming module which is capable of forming a component of an integrated multiregional model. In section 4.2 an outline is given of the structure of interregional programming models in both economy-wide and individual sector applications. In section 4.3 we discuss the integration of multiregional linear programming and input-output modules. More specifically, we discuss:

(a) how inclusion of input-output relations as constraints in appropriately designed economy-wide interregional programming models permit (i) levels of regional investment (capacity expansion) in each sector to be derived endogenously instead of being consolidated into an endogenously determined set of final demands, and (ii) levels of regional activity in each sector to be derived in a manner consistent with the policy goals (ambitions) of different sectors, regions and/or levels of government; and

(b) how regional production/trade patterns in one sector (say energy or agriculture) could be projected via use of an interregional programming model, while retaining the use of an interregional input-output model for projecting production/trade patterns for other sectors of the multiregional economy.

In Section 4.4 the integration of comparative cost-industrial complex, programming and input-output modules is presented as a sequence of steps. Finally, some concluding remarks are made in Section 4.5.

4.2 Outline of Multiregional Linear Programming Models

Interregional linear programming recognizes that within any given multiregion system there are many feasible production alternatives within each region and/or sector and many possible trade flow alternatives between regions and/or sectors. It seeks to identify that set of production and trade flow plans which either maximizes a 'desired' benefit or minimizes an 'undesired' loss, subject to a set of constraints reflecting regional resource and factor endowments, regional production technologies, national and/or regional socio-political goals, and so on. The approach is extremely flexible in that there are many possible objectives (or combinations of objectives) which could be optimized, and many sets of constraints which could be imposed on the results. For example, the objective function may relate to:

(a) a national policy objective (or combination of national objectives), such as maximize gross national product or minimize national energy consumption, or minimize a weighted sum of national unemployment and national energy consumption; or

(b) a particular region's policy objective (or combination of objectives), such as maximize regional per capita income, or minimize regional unemployment; or

(c) some combination of national and regional policy objectives.

Similarly, the constraints may embody resource, technological, and socio-political constraints relating to each region, or combination of regions, or to the nation as a whole. However, use of this approach does require that the program be able to be stated in the form of a linear objective function subject to a set of linear inequalities. This restricts the applicability of the approach, since it cannot deal with such realistic phenomena as scale economies, agglomeration economies, or nonlinear interaction among objectives.[1]

4.2.1 Structure of a General Interregional Programming Model

We first present a general interregional linear programming model which maximizes the sum of regional incomes subject to resource, technological, minimum consumption level, and non-negativity constraints.[2] The model is non-operational in the form presented, but does serve to illustrate the basic structure of a multiregional programming model.

The nation is divided into U regions (J=A,...,U). Each region is assumed to contain Â' activities (Â=1, ...,Â'), and produce and/or consume m commodities (h=1,...,m).

The first k commodities (h=1,...,k) are taken to be either <u>resources</u> or <u>intermediate products.</u> Each resource may be either mobile or immobile, however, its initial endowment (supply) in each region is assumed to be fixed and known beforehand. The remaining (m-k) commodities (h=k+1,...,m) are <u>finished products</u>. Since some items can be consumed either as intermediates (by industries) or finished products (e.g., by households), these items are treated as two commodities. That is, as an intermediate product in the first k commodities and as a finished product in the remaining (m-k) commodities. These items are designated as h=r+1,...,m in the latter product listing, while items which can be viewed as pure finished products are designated as h=k+1,...,r.

The first n activities (Â=1,...,n) are <u>productive activities</u>, each one of which employs resources and/or intermediate products as inputs and yields finished and/or intermediate products as outputs. The second set of k activities (Â=1,...,k) are <u>dummy activities</u>, each one of which transforms a single input comprising either an intermediate product or a resource into a single equivalent output in the form of a finished product.[3] The final set of Â' - (n+k) activities are <u>trade flow activities</u> which are responsible for transporting each of the commodities h=1,...,r among the U regions.[4] Thus, any given region J may have as many as (U-1)r trade flow activities,[5] each of which is associated with a single output (namely, the commodity received as an import in the region of destination) and generally require at least two inputs (namely, the commodity being exported from the region of origin, and transportation).[6]

The objective function for the model can be expressed as:

$$\max Z = \sum_{J=A}^{U} \sum_{\hat{A}=1}^{\hat{A}'} {}^J c_{\hat{A}} \, {}^J x_{\hat{A}}$$

(4.2.1)

where ${}^J x_{\hat{A}}$ = the level at which activity $\hat{A}$ is operated in region J;

${}^J c_{\hat{A}}$ = the income generated in region J per unit operation of activity $\hat{A}$ in J; and

Z = the sum of regional incomes for the nation (system).

The model may be subject to as many as Um constraints concerned with the use of resources in each region; the input-output balance re: intermediate products in each region; and the availability of finished products for consumption by households in each region.

The general form of these constraints can be expressed as:

$$-\sum_{\hat{A}=1}^{\hat{A}'} {}^J a_{h\hat{A}} \, {}^J x_{\hat{A}} - \sum_{\substack{L=A \\ L \neq J}}^{U} {}^{LJ} a_{h\hat{A}} \, {}^L x_{\hat{A}} \leq {}^J R_h$$

$(h=1,\ldots,m;\ J=A,\ldots,U)$

(4.2.2)

or equivalently as:

$$-\sum_{L=A}^{U} \sum_{\hat{A}=1}^{\hat{A}'} {}^{LJ} a_{h\hat{A}} \, {}^L x_{\hat{A}} \leq {}^J R_h$$

(4.2.3)

where $^J R_h$ = the limiting magnitude pertaining to commodity h. Where h is a resource, $^J R_h$ is the fixed supply (endowment) of resource h in region J; where h is an intermediate product, $^J R_h$ equals zero; and where h is a finished product, $^J R_h$ is the fixed quantity of finished product h which must be available for consumption by households in region J;

$^J a_{h\hat{A}} = ^{JJ} a_{h\hat{A}}$ = the input requirements of commodity h per unit of activity $\hat{A}$ in region J when negative, and the output of commodity h per unit of activity $\hat{A}$ in region J when positive;[7] and

$^{LJ} a_{h\hat{A}}$ (L≠J)= the level of shipment (output) of commodity h from region L to region J per unit of activity $\hat{A}$ in L when positive; and when negative the input requirements of commodity h per unit of such a trade flow activity $\hat{A}$.

Thus, where h is an immobile resource in J, $^{JL} a_{h\hat{A}} \neq 0$ for all $\hat{A}$, L, and equation (4.2.2) states that the sum of requirements for this resource (i) as an intermediate input by all productive activities in region J, (ii) as an input to be transformed into a finished product by the appropriate dummy activity in region J, and (iii) as a local intermediate input by all trade flow activities in region J, must not be greater than (iv) its available (endowment) supply in region J.

Where h is a mobile resource in J, $^{JL} a_{h\hat{A}} \neq 0$ for some $\hat{A}$,L. Thus, equation (4.2.2) now states that (i) requirements of this resource as a local input by all activities, plus (ii) requirements of it by trade flow activities for exports to other regions, less (iii) total imports of it from other regions' trade flow activities, must not be greater than (iv) the given local supply (endowment) of that resource.

Where h is an intermediate product, equation (4.2.2) states that the requirements for this product (i) as an intermediate input to all productive activities in region J, (ii) as an input to be transformed into a finished product by the appropriate dummy activity in J; (iii) as a local intermediate input and as an input to be exported by all trade flow activities in region J, less (iv) the output of intermediate product h by all productive activities in J, less (v) imports of intermediate product h from all other regions L(L=A,...,U), must not be greater than zero.

Finally, where h is a finished product, equation (4.2.2) states[8] that the sum of (i) outputs of finished product h by all productive and dummy activities in region J, less (ii) exports of finished product h via trade flow activities to all other regions, must not be less than (iii) the fixed quantity of finished product h which must be available for household consumption in region J.

The model is also subject to non-negativity constraints for each activity $\hat{A}$, that is:

$$^J x_{\hat{A}} \geq 0 \qquad (J=A,\ldots,U; \quad \hat{A}=1,\ldots,\hat{A}') \qquad (4.2.4)$$

From the above definitions it is clear that, given a set of finished product prices $^J p_h$ (h=k+1,...,m;J=A,...,U), the set of coefficients $^J c_{\hat{A}}$ ($\hat{A}$=1,...,$\hat{A}$';J=A,...,U) required for the objective function given in equation (4.2.1) can be derived as:

$$^J c_{\hat{A}} = \sum_{h=k+1}^{m} {}^J a_{h\hat{A}} \, {}^J p_h \qquad (4.2.5)$$

where $^J a_{h\hat{A}}$ (h = k+1, ..., m) are the unit level ouputs of finished products by activity $\hat{A}$ in region J.[9]

When there are many activities, commodities and/or regions involved, the above interregional program may become non-operational (ie., incapable of yielding a numerical solution, given data availability and computational facilities). In any given application, the model can be scaled down to a manageable size by:

(a) restricting its application to one or a few sectors of the multiregional economy;

(b) adopting a highly aggregate system of regions, commodities and/or activities when focussing on the entire multiregional economy; or

(c) restricting its application to a subset of commodities/activities and use of supplementary modes of analysis (such as input-output, comparative cost and/or industrial complex approaches) to project activity levels within the rest of the multiregional economy.

In the next section, some applications which focus on only a small subset of commodities/activities will be presented. Other more general applications of the interregional programming approach will be discussed in section 4.3.1.

4.2.2 Individual Sector Interregional Programming Models

Henderson (1958) employed an interregional programming model to identify the most efficient pattern of interregional coal shipments within the United States during 1947. The nation was divided into 14 regions (of which 11 had the capacity to produce coal), and because of differences in extraction costs two types of mining operations (underground and surface) were distinguished when defining activities. The objective function selected involved minimization of the total cost C of extracting (producing) and delivering (transporting) coal for the system. That is:

$$\min C = \sum_{\hat{A}=1}^{2} \sum_{J=A}^{U} \sum_{L=A}^{U} {}^{JL}c_{\hat{A}} \, {}^{JL}x_{\hat{A}}$$

(4.2.6)

where ${}^{JL}x_{\hat{A}}$ = number of units of coal extracted via mining operations of type Â in region J and delivered to consumers in region L;

$^{JL}c_{\hat{A}}$ = average cost of extracting a unit of coal via mining operations of type Â in region J and delivering it to consumers in region L. Regardless of the level of $^{JL}X_{\hat{A}}$, this cost is assumed to remain constant;

Â = a subscript designating the type of mining operation. When Â =1, we have an underground mine, and when Â =2, we have a surface mine.

A first set of constraints was introduced to ensure that for each type of mining operation Â and each producing region J, the sum of coal shipments from that region does not exceed the available capacity. That is:

$$\sum_{L=A}^{U} {}^{JL}X_{\hat{A}} \leq {}^{J}R_{\hat{A}} \qquad (4.2.7)$$

where $^{J}R_{\hat{A}}$ = the total capacity of region J for mining operations of type Â.[10]

A second set of constraints was introduced to ensure that for each consuming region L, the sum of coal shipments to that region is at least as great as the demand for coal in that region. That is:

$$\sum_{J=A}^{U} \sum_{\hat{A}=1}^{2} {}^{JL}X_{\hat{A}} \geq {}^{L}D \qquad (4.2.8)$$

The final set of constraints required that coal shipments $^{JL}X_{\hat{A}}$ for all pairs of regions (J,L=A,...,U) and activity-types (Â =1, 2) be nonnegative.

The operation of the Henderson (1958) model requires exogenous estimates of regional demands D^L, regional capacities $^JR_{\hat{A}}$, and unit extraction and transportation costs $^{JL}c_{\hat{A}}$. It then yields the pattern of shipments $^{JL}X_{\hat{A}}$ and production levels $^JX_{\hat{A}}$ (= $\Sigma\; ^{JL}X_{\hat{A}}$) which is most cost efficient under these conditions.[11]

A much more elaborate interregional programming model has been developed for the energy sector of the United States by the Brookhaven National Laboratory (see Goettle (1977), Goettle et al (1977) and Cherniavsky (1974)). The model starts from the extraction of energy resources r (r = 1, ..., 9), taking into account the relevant supply schedules in each of the 9 census regions.[12] These resources are then transformed into a more marketable set of products, designated IEF's or intermediate energy forms e (e = 1, ..., 10), taking into account different possible conversion technologies.[13] These intermediate energy forms are then used to satisfy an exogenously given demand for energy, distributed over functional energy services d (d = 1, .., 17).[14] In ensuring that any given functional energy service is provided at the required level, the possibility of producing it from a number of different combinations of IEF's is considered.

Currently, the objective function of the model consists of minimizing the total costs involved in extraction and transportation of energy resources, processing resources into intermediate energy forms, transportaton of intermediate energy forms and conversion of intermediate energy forms into functional energy services. Figure 4.1 gives a schematic representation of the basic structure of the model.

Although the two individual sector studies presented in this section focus on the same sector - namely energy - of the multiregional economy, other applications have focussed on other sectors. Of particular interest in the Australian context are those applications which focus on the agricultural sector (see, for example, Heady (1964), Snodgrass and French (1958) and Henderson (1958)). Appendix 4A contains further relevant discussion.

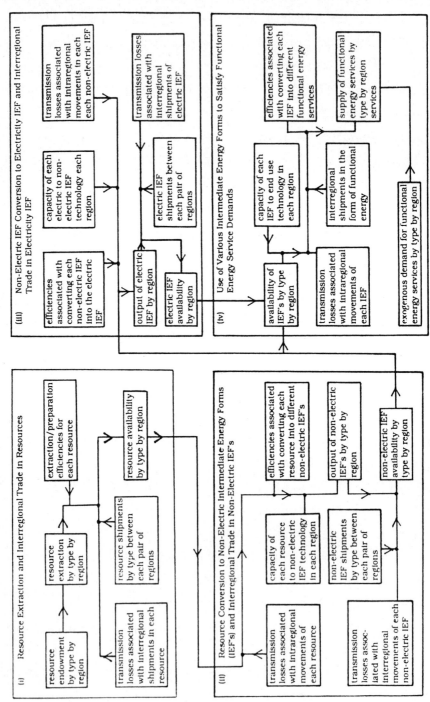

Figure 4.1 A Schematic Representation of Paths Between Energy Resource Supplies and Functional Energy Service Demand Satisfaction in the Brookhaven Programming Model

Regardless of which sector provides the focus for an individual sector interregional programming model, this model must be supplemented by the use of other techniques if the analyst is interested in the entire multiregional economy. See pp. 121-124 for further discussion.

4.3 Integration of Multiregional Linear Programming and Input-Output Models

4.3.1 Input-Output Relations as Constraints in Programming Models

In constrast to the individual sector studies just discussed, another approach to operationalizing the general interregional programming model retains a focus on the entire multiregional economy. This approach generally divides the economy into a relatively small number of regions and/or activities. Each 'activity' is assumed to produce only one output and hence becomes a 'sector' or 'industry' in the conventional input-output sense. As a consequence, the set of constraints summarized by equation (4.2.3) are divided into three sets, namely resource use, consumption level, and input-output type balance relations. The third set of relations are then replaced by balance equations of the form:

$$^J X_i + \sum_{\substack{L=A \\ L \neq J}}^{U} \sum_{j=1}^{n} {}^{LJ}\hat{a}_{ij} {}^J X_j - \sum_{j=1}^{n} {}^{JJ}\hat{a}_{ij} {}^J X_j$$

$$- \sum_{\substack{L=A \\ L \neq J}}^{U} \sum_{j=1}^{n} {}^{JL}\hat{a}_{ij} {}^L X_j \geq {}^J Y_i$$

$$(i=1,\ldots,m; J=A,\ldots,U) \qquad (4.3.1)$$

associated with the Isard ideal-type of interregional input-output model discussed on pp. 17-22 (or of an equivalent form when one of the alternatives to this model is employed).

A simple example of such a model is given by Hurter and Moses (1964).[15] The objective function selected was the minimization of the sum of transport costs plus the direct costs of using primary inputs (land, labor, capital, etc.). These primary inputs were assumed to be interregionally immobile, so that this objective function can be written as:

$$\min c = \sum_{J=A}^{U} \sum_{L=A}^{U} \sum_{i=1}^{n} [(^{JL}p_i)(^{JL}X_i)]$$

$$+ \sum_{J=A}^{U} \sum_{i=1}^{n} \sum_{r=1}^{r'} [(^{J}p_r)(^{J}a_{ri})(^{J}X_i)] \quad (4.3.2)$$

where $^{JL}p_i$ = cost of transporting one unit of sector i's output from region J to L;

$^{J}p_r$ = cost (price) of one unit of primary input r in region J;

$^{J}a_{ri}$ = number of units of primary input r required per unit of sector i's output.

In addition to the usual non-negativity constraints and the Un input-output balance constraints of the form given in equation (4.3.1), other constraints placed on the model required that for each region J:

(a) the level of output $^{J}X_i$ in each sector does not exceed the available productive capacity $^{J}R_i$, and

(b) the total use of each primary input r, that is $_i\Sigma\ ^{J}a_{ri}\ ^{J}X_i$ does not exceed the available supply $^{J}R_r$ of that input.

More complex applications of this type of approach are possible, since the use of a combined input-output/programming model facilitates the introduction of some dynamic elements into the balance equations characterizing interregional input-output relations.[16] In particular the levels of investment (capacity expansion) in each sector can be derived endogenously instead of being consolidated into an exogenously determined set of final demands. The basic additional ingredient to the Isard ideal-type interregional input-output structure is a (Un x Un) matrix of capital coefficients B, with typical element $^{JL}b_{ij}$ which states the amount of output of sector i in region J required per unit of capacity expansion in sector j in region L. Ignoring the possibility of unused production capacity,[17] we have the following input-output relations for each point of time t ($t=t,\ldots,t+\theta$), sector i (i = 1, ..., n) and region J (J=A, .., U):

$$^{J}X_{it} + \sum_{\substack{L=A \\ L \neq J}}^{U} \sum_{j=1}^{n} {}^{LJ}\hat{a}_{ijt}\, {}^{J}X_{jt} + \sum_{\substack{L=A \\ L \neq J}}^{U} \sum_{j=1}^{n} {}^{JL}b_{ijt}\, {}^{J}I_{jt} \geq$$

$$\sum_{L=A}^{U} \sum_{j=1}^{n} {}^{JL}\hat{a}_{ijt}\, {}^{L}X_{jt} + \sum_{L=A}^{U} \sum_{j=1}^{n} {}^{JL}b_{ijt}\, {}^{L}I_{jt} + {}^{J}Y_{it}$$

(4.3.3)

where $^{J}I_{it}$ = the induced expansion in capacity of sector i in region J at time t; and

$^{J}Y_{it}$ = the final demand for the output of sector i in region J at time t now excludes investment demands.

The coefficients $^{JL}\hat{a}_{ijt}$ and $^{LJ}b_{ijt}$ each have subscripts t to allow for the possibility of changes during the projection period due to technological change, etc. Thus, the set of equations represented by (4.3.3) state that for each region J and time period t, the total supply of sector i's output available in region J (given by the LHS of the inequality) must not fall short of the total demands for that output (given by the RHS of the inequality).

For computational purposes, the induced expansions in capacities ($^J I_{jt}$) must be related to the output levels $^J X_{jt}$. One way this can be done, ignoring the need to consider the effects of depreciation,[18] is to introduce another set of equations:

$$^J K_{jt} = {^J K_{j0}} + {^J I_{jt}} + \ldots + {^J I_{jt}} \geq {^J k_{jt}} \, {^J X_{jt}} \quad (J=A, \ldots, U; j=1, \ldots, n)$$
(4.3.4)

where $^J K_{jt}$ = capacity (capital stock) of sector j in region J at the end of time period t;

$^J K_{j0}$ = existing capacity at beginning of time period t; and

$^J k_{jt}$ = capacity (capital stock) requirements per unit output of sector i in region J at time t.

Thus the set of equations represented by (4.3.4) state that for each region J, time period t and sector j, the level of capital expansion $^J I_{jt}$ must be such that the resulting capacity available (given by the LHS) is not less than the total capacity required (given by the RHS).

Applications of this type of model are non-existent because of its heavy data requirements. However, two models, both by Swedish scholars, which resemble it most closely are discussed below.

a. The REGAL (REGional ALlocation of all productive sectors) model developed by Granholm and Snickars (1979) and Snickars and Granholm (1981) simplifies the above model in the following manner:

(i) a one time period projection is all that is made, so that the time subscripts t in equations (4.3.3) and (4.3.4) can be suppressed;

(ii) all requirements of sector i's output for capital expansion in any given region J are assumed to be purchased from the local pool of that sector's output, so that the (Un x Un) matrix of B coefficients underlying equation (4.3.3) is replaced by a set of U matrices, designated B, of order (n x n) with typical element $^J b_{ij}$ stating for region J the amount of sector i's output required per unit of capital expansion in sector j;

(iii) the Chenery-Moses type trade coefficients $^{JL}c_i$ were employed instead of the more detailed $^{JL}c_{ij}$ coefficients employed in the Isard ideal-type model, so that the $^{JL}\hat{a}_{ij}$ coefficient in equation (4.3.3) is replaced by expression $(^{JL}c_i)$ $(^J a_{ij})$;[19] and

(iv) the technical coefficients $(^J a_{ij})$ and capital coefficients $(^J b_{ij})$ were assumed to be identical to the corresponding national coefficients for all regions J.

Within the limitations imposed by these simplifying assumptions, the aim of the REGAL model is to consider explicity the role that public sector infrastructure provision can play in regional development planning.

The Swedish economy is divided into 8 regions and 22 industrial sectors, and projections of national output, employment, population and capacity requirements are determined via the use of a national input-output/econometric model. For 6 sectors, these national magnitudes are allocated among regions using predetermined (i.e., exogenous) share coefficients.[20] In the remaining 16 sectors, these magnitudes are allocated among regions in such a way as to minimize the total cost C of public and private sector investments. That is:

$$\min C = \sum_{J=1}^{8} \sum_{i=1}^{22} {}^J c_i \, {}^J I_i$$

(4.3.5)

where $^J c_i$ = unit cost of investments (i.e., capacity expansions) in sector i in region J;

^JI_i = level of investments (capacity expansions) in sector i in region J.

The constraints imposed on the model consist of three fundamental types:

(i) a set of technology descriptions wherein each region's economy is characterized by linear relations;

(ii) a set of restrictions on the national and/or regional levels of the magnitudes (output, employment, population, and capital stock) being allocated; and

(iii) a set of politically determined supply/demand (minimum requirement) restrictions on the level of the public sector infrastructure provision in each region.

In addition to the input-output balance equations of the form just discussed, other technological constraints act to:

(i) specify for each sector in any given region a linear relationship between output levels and value added, and between value added and the requirements for capital and labor inputs; and

(ii) define capital stock (i.e., capacity levels) in any given region at the end of the projection period as the sum of depreciated initial capital stock (capacity) plus the new investment (capital expansion) conducted during that period.[21]

Since the model is concerned with the regional allocation of given national (endogenous sector) forecasts of output, employment, population and capacity expansions, one set of restrictions in the second category of constraints ensures that the sum over all regions of these magnitudes does not fall short of its corresponding national target (forecast) level. For example, for employment levels in these sectors we have:

$$\sum_{J=1}^{8} {}^{J}L_i \geq L_i \qquad i=1,\ldots,16 \qquad (4.3.6)$$

where L_i is the forecasted national employment level in sector i; and ${}^{J}L_i$ is the employment level in sector i in region J.

For those sectors whose regional allocation is determined exogenously, we have a set of equality constraints for each of the above magnitudes. For example, for employment levels we have a set of constraints of the form:

$$ {}^{J}L_i = {}^{J}\lambda_i L_i \qquad i=17,\ldots,22 \qquad (4.3.7)$$

where ${}^{J}\lambda_i$ is the proportion of the forecasted national employment level in sector i that is allocated to region J.

Finally, in order to ensure that the model does not produce unreasonable shifts in economic activity, upper and lower bound restrictions are placed on the regional levels of each of the above magnitudes. For example, for employment levels we have:

$$ {}^{J}\underline{L} \leq {}^{J}L \leq {}^{J}L' \qquad (4.3.8)$$

where ${}^{J}L'$ and ${}^{J}\underline{L}$ are upper and lower limits imposed on total employment in region J, respectively.

The final set of constraints imposes limitations on public sector technical, economic and social infrastructure in each region on the basis of a politically determined minimum standard. For example, for technical and economic infrastructure we have:

$$\sum_{i \in I_{pu}} {}^J K_{ij} \geq {}^J \lambda_j \left[\sum_{i \in I_{pr}} {}^J K_i \right]$$

(4.3.9)

where ${}^J K_{ij}$ = provision of technical and economic infrastructure of type j by public sector i in region J;

${}^J K_i$ = capital stock (i.e., capacity levels) in private sector i in region J;

${}^J \lambda_j$ = a parameter reflecting the minimum requirement for infrastructure of type j per unit level of capital stock in all private sectors in region J; and

I_{pu} and I_{pr} = public and private sector investment sectors respectively.

For each type of social infrastructure, the model identifies a politically acceptable access measure in relation to some target population group and introduces these measures as minimum requirements. That is:

$$\sum_i {}^J K_{i\ddot{E}} \geq {}^J \Omega_{\ddot{E}} \, {}^J \lambda_{\ddot{E}} \, {}^J \underline{P} \qquad i \in I_{pu}$$

(4.3.10)

where ${}^J \lambda_{\ddot{E}}$ = proportion of $\underline{P}^J$, region J's population, falling within the target group identified as primary users of infrastructure of type Ë; and

${}^J \Omega_{\ddot{E}}$ = a parameter reflecting the minimum requirement for social infrastructure of type Ë per member of the corresponding target population group in region J.

Although data limitations force the results of this model to be fairly crude, it is nevertheless useful for examining the tradeoffs between minimum requirement levels set for the different types of public infrastructures, and between target levels set for the different regions regarding employment, output, population, etc. However, this points up another problem to be confronted when employing a multiregional programming model. That is, when non-technical types of constraints are imposed on the model, how should the corresponding binding magnitudes be determined? This problem is taken up in the discussion of possible INPOL (Interdependent Policy Formation) module in Isard and Smith (1982 and 1984a).[22]

b. The MORSE (MOdel for the analysis of Regional development, Scarce resources and Employment) model developed by Lundqvist (1980, 1980a, and 1981) provides another example of a combined input-output/programming model incorporating dynamic aspects of capital formation. A schematic representation of the model is given in Figure 4.2. The aims of MORSE are:

(i) to identify regional (and hence, national) development patterns consistent with minimum requirements for national economic, employment and energy policies; and

(ii) given that such feasible development patterns do exist, to select the pattern that is best in terms of a given objective (or combination of objectives).

The Swedish economy is divided into 8 regions and 9 industrial sectors, and the national multisectoral growth model (MSG) developed by Bergman and Por (1980) is used to project national trends in output and primary factor (capital, labor and energy) requirements for each sector, the composition of foreign trade, and household consumption expenditure patterns. For those sectors whose location patterns are determined primarily by the geographic distribution of resources,[23] the national output levels projected by MSG are allocated among the 8 regions using predetermined (exogenous) share coefficients. For the remaining sectors, regional output levels (and subsequently a host of other variables, including consumption, investment, employment, interregional and foreign trade and energy usage) are determined in such a way as to maximize the discounted sum of regional volumes of consumption. That is:

CHAPTER 4

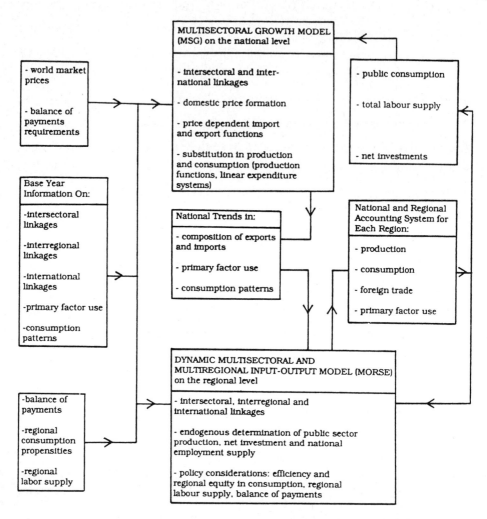

Figure 4.2 Structure of the MORSE Model

$$\max \sum_{t=t}^{t+\theta} \varepsilon_t \sum_{J=A}^{U} {}^J Q_t \qquad (4.3.11)$$

where ${}^J Q_t$ = consumption level in region J at time t;

ε_τ = rate of time preference (i.e., discount rate) re: consumption. The constraints imposed on the model consist of three fundamental types: commodity balance, supply and political preference.[24]

The commodity balance constraints employed in MORSE approximate those in equation (4.3.3) more closely than the REGAL model just discussed. However, the MORSE model is closed with respect to households, which means that the ${}^J Y_{it}$ vector now excludes both household consumption and investment demands. In addition, the following simplifications are made:

(i) the technical-trade coefficients adopted employ a Chenery-Moses type of assumption, so that again the ${}^{JL}\hat{a}_{ij}$ coefficient in (4.3.3) is in effect replaced by the expression $({}^{JL}c_i)({}^J a_{ij})$;[25]

(ii) the capital coefficients adopted also employ a Chenery-Moses type of assumption, so that the ${}^{JL}b_{ij}$ coefficients in (4.3.3) is in effect replaced by the expression $({}^{JL}g_i)\,({}^J b_{ij})$ where ${}^{JL}g_i$ = the proportion of region L's purchases of sector i's output for capital expansion purposes which comes from region J;[26]

(iii) each region's ${}^J b_{ij}$ coefficients are assumed to be identical to the corresponding national coefficients estimated via the use of the MSG model; and

(iv) with minor exceptions,[27] each region's technical and capital coefficients (${}^J a_{ij}$ and ${}^J b_{ij}$) and interregional trade coefficients (${}^{JL}c_i$ and ${}^{JL}g_i$) are assumed to be constant over the projection period.[28] As a result, the time subscript t is generally suppressed on these coefficients.

The <u>supply constraints</u> relate to the factors of production (resources) in scarce supply, namely labor, capital and energy. For example for each region J and time period t, the labor constraints require, that:

$$\sum_{i=1}^{9} {}^J\beta_{it} \, {}^JX_{it} < {}^JL'_{it}$$

(4.3.12)

where ${}^J\beta_{it}$ = labor input required per unit of output ${}^JX_{it}$ of sector i in region J at time t; and

${}^JL'_{it}$ = exogenously determined supply of labor in region J at time t.

The time paths of sectoral labor-output coefficients ${}^J\beta_{it}$ (i = 1, ..., 9) are derived from the national MGS model, and it is assumed that each region's coefficients follow the same time path as their national counterpart.[29] The energy constraints are formulated at the national level and require, for each energy type m (m = 1, ..., 4) and time period t, that:

$$\sum_{J=1}^{8} \left[\sum_{i=1}^{9} {}^Je_{mit} \, {}^JX_{it} + {}^Je_{mqt} \, {}^JQ_t \right] < E'_{mt}$$

(4.3.13)

where ${}^Je_{mit}$ = input of energy type m required per unit output X_{it} of sector i in region J at time t; and

${}^Je_{mqt}$ = input of energy type m required per unit of household consumption ${}^JQ_t$ in region J at time t.

Finally, for any given region and time period the capital constraints state that the production capacity (i.e., capital inputs) required by each sector cannot exceed the currently available capital stock (i.e. the sum of depreciated initial capital stock and depreciated investments carried out in earlier time periods).[30]

The political constraints have been imposed to ensure that certain socio-economic goals are satisfied in each time period. That is:

(i) to satisfy the goal of regional equity, lower bounds are placed on both the proportion of gross domestic product devoted to household consumption in each region, and the total employment level in each region;

(ii) to satisfy the goal of high national economic growth, a lower bound is placed on capital formation when summed over all sectors and regions; and

(iii) to satisfy the goal of favorable national balance of payments, a lower bound is placed on the net surplus of foreign trade when summed over all sectors and regions.

The MORSE model has been implemented for three consecutive 5-year periods (1975-1990), and used to study the sensitivity of regional development patterns with respect to different (a) national energy policy scenarios, (b) rates of time preference (ε_t) re: consumption, and (c) rates of regional self-sufficiency in the output of certain sectors. In addition, experiments have been conducted with the use of other objective functions such as:

(i) minimization of some aggregate of national energy use,

$$\min \sum_{t=t}^{t+\theta} \sum_{J=A}^{8} \sum_{m=1}^{4} \mu_m \left[\sum_{i=1}^{9} {}^J e_{mit} {}^J X_{it} + {}^J e_{mqt} {}^J Q_t \right] \quad (4.3.14)$$

where μ_m = an exogenously determined weight attached to energy type m;

(ii) maximization of national employment,

$$\max \sum_{t=t}^{t+\theta} \sum_{J=A}^{U} \sum_{i=1}^{8} {}^J \beta_{it} {}^J X_{it} \quad (4.3.15)$$

(iii) maximization of a weighted sum of the consumption, energy and employment objectives given in equations (4.3.11) (4.3.14) and (4.3.15), respectively.

Then by changing (a) the binding magnitudes on the achievement of the different policy goals incorporated in the various constraints, and (b) the weights attached to the policy goals incorporated in the objective function, the nature of the tradeoffs (conflicts) between the different policy ambitions (goals) have been identified. However, this points up another problem to be confronted when employing a multiregional programming model. That is, (a) which policy objectives (goals) should be placed in the objective function and which should be introduced in the form of constraints, and (b) which set of weights should be employed when the objective function relates to more than one policy objective (goal). This problem is taken up in the discussion of a possible INPOL (Interdependent Policy Formation) module in Isard and C. Smith (1982, and 1984a).[31]

Other examples of models employing a combined multiregional programming and dynamic input-output model can be found in the literature.[32] However, the two examples presented above point up clearly the dilemma confronting the analyst --- the introduction of the time dimension can increase the scale of the model enormously for any reasonable level of regional/sectoral disaggregation. Hence, given current data availability and/or regional budgets, the analyst is often forced to sacrifice:

(i) region-specific $^{JL}\hat{a}_{ij}$ and/or $^{JL}b_{ij}$ coefficients, as in the REGAL model; and/or

(ii) sectoral detail, which when taken to the extreme (as with the 9 endogenous sectors in the MORSE model) brings into serious question the validity of the input-output assumption that each sector employs a single production technology and produces a homogenous set of outputs.

Clearly, before deciding whether to adopt a dynamic rather than a static-type input-output model, the question that needs to be answered is whether the added complexity of the dynamic model is really worth it given the nature of the tradeoffs involved.

4.3.2 Linkage of Multiregional Linear Programming and Input-Output Modules

As discussed above when a multiregional programming model is employed to examine the optimal production/trade pattern within an industrial sector (or set of interrelated sectors) the model must be supplemented by the use of other techniques if the analyst is interested in the entire multi-regional economy. One of the techniques which is especially useful is the interregional input-output model discussed on pp.17-22. The advantages of this formulation are:

(i) the linear programming submodel can allow substitution between inputs and variable interregional trade patterns for the 'optimizing' sector(s) -- while this flexibility is precluded in the standard interregional input-output submodel; and

(ii) the input-output submodel allows interaction of the 'optimizing' sector(s) with the rest of the interregional economic system to be taken into account explicitly.

The best example of such a model is provided by the Brookhaven (BNL) Multiregional Energy and Interindustry Model of the United States developed by Goettle et al (1977), and extended by Isard and Anselin (1980). The linear programming component of this model was presented on pp. 105-106. The implementation of the interregional input-output component involves dividing each region's economy into three sets of sectors: energy supply sectors S, energy product sectors P, and non-energy sectors N.[33] The interregional technical-trade coefficients matrix Æ can then be partitioned as follows:

$$\mathcal{E} = \begin{bmatrix} \mathcal{E}_{SS} & \mathcal{E}_{SP} & 0 \\ \mathcal{E}_{PS} & 0 & \mathcal{E}_{PN} \\ \mathcal{E}_{NS} & 0 & \mathcal{E}_{NN} \end{bmatrix} \qquad (4.3.16)$$

The energy product sectors P are dummy sectors introduced to convert energy resources produced by the energy supply sectors into energy end-products required to satisfy locally:

(i) intermediate demands from the energy supply sectors S (reflected in the set of $\mathcal{E}_{PS}$ coefficients) and the non-energy sectors N (reflected in the set of $\mathcal{E}_{PN}$ coefficients);[34] and

(ii) final demands for energy end-products (summarized in a vector Y_P). The only inputs employed by these sectors are energy resources, reflected in the set of $\mathcal{E}_{SP}$ coefficients. As a consequence the submatrices $\mathcal{E}_{PP}$ and $\mathcal{E}_{NP}$ are zero.

The non-energy sectors N employ inputs from the energy product sectors P and themselves, reflected in the sets of $\mathcal{E}_{PN}$ and $\mathcal{E}_{NN}$ coefficients, respectively.[35] The outputs of these sectors are non-energy products required to satisfy:

(i) intermediate demands from energy supply sectors in all regions (reflected in the set of $\mathcal{E}_{NS}$ coefficients) and non-energy sectors in all regions (reflected in the set of $\mathcal{E}_{NN}$ coefficients); and

(ii) final demands for non-energy products from all regions (summarized in a vector Y_N).

The energy supply sectors S are responsible for extracting and/or processing the energy resources required by the energy product sectors P. In so doing, they employ inputs from the energy product and non-energy sectors as well as from themselves. These input requirements are reflected in the $\mathcal{E}_{PS}$, $\mathcal{E}_{NS}$, and $\mathcal{E}_{SS}$ sets of coefficients, respectively. By convention, the only final demand for the outputs of the energy supply sectors (summarized in the vector Y_S) are foreign exports and inventory.[36]

This particular sectoral structure suggests an iterative solution procedure[37] (depicted in Figure 4.3) comprising the following steps:

Step 1: *Derivation of Initial Final Demand Projections*

The first step involves the determination of (a) those interregional technical-trade coefficients ($Æ_{PS}$, $Æ_{PN}$, $Æ_{NS}$, and $Æ_{NN}$) which are invariant with respect to the substitution processes embodied in the linear-programming (LP) submodel; (b) those regional cost parameters, conversion and transmission efficiencies, capacity limits, etc. (required for the operation of the LP submodel) which are invariant to the output levels determined in the input-output (I-O) submodel; and (c) those final demand components (Y_N, Y_P, and Y_S) invariant to both the LP and I-O submodel solutions.

Step 2: *Derivation of Initial Output Projections for Non-Energy Sectors*

Given the final demands Y_N for non-energy sector outputs, the direct and indirect output requirements can be calculated as:

$$X_N = (I - Æ_{NN})^{-1} Y_N \qquad (4.3.17)$$

Step 3: *Derivation of Initial Output Projections for Energy Product Sectors*

The direct and indirect output requirements for the energy product sectors can then be calculated as:

$$X_P = Æ_{PN} X_N + Y_P \qquad (4.3.18)$$

Step 4: *Projection of Functional Energy Service Demand Level*

The output estimates X_P derived in step 3 provide a first approximation of regional demands for energy end-products. By

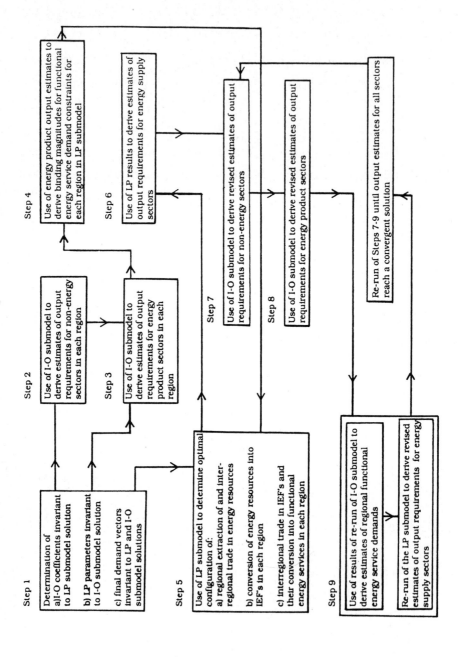

Figure 4.3 Linkage of Multiregional Linear Programming and Input-Output Submodels

employing a set of simple aggregation and/or disaggregation procedures,[38] these estimates can be converted into the functional energy service demand levels $^{J}D_d$ (d = 1, ..., 17) required as binding magnitudes for the demand equations in the linear programming submodel.

Step 5: *Initial Run of Linear Programming Submodel*

The linear programming submodel is run to determine an optimal configuration of (a) regional extraction of and interregional trade in energy resources (r = 1, ..., 9); (b) conversion of energy resources into intermediate energy forms (e = 1, ..., 10) in each region; and (c) interregional trade in intermediate energy forms and their conversion into functional energy services (d = 1, ..., 17) in each region.

Step 6: *Derivation of Initial Output Projections for Energy Supply Sectors*

The results derived in Step 5 are used to provide a first approximation to the output X_S of energy supply sectors, as well as the $Æ_{SS}$ and $Æ_{PS}$ submatrices of (4.3.16).[39]

Step 7: *Derivation of Revised Output Projections for Non-Energy Sectors*

The results derived in Step 6 are used to produce a revised projection of the output of non-energy sectors. That is:

$$X_N = (I - Æ_{NN})^{-1} (Y_N + Æ_{NS} X_S) \qquad (4.3.19)$$

Step 8: *Derivation of Revised Output Projections for Energy Product Sectors*

The results derived in Steps 6 and 7 are used to produce a revised projection of the output of energy product sectors. That is:

$$X_P = \mathcal{E}_{PN} X_N + Y_P + \mathcal{E}_{PS} X_S \qquad (4.3.20)$$

Step 9: *Successive Reruns of Input-Output and Programming Modules*

The results derived in Step 8 are used to provide revised estimates $^J D_d$ of regional demand for functional energy services. The linear programming submodel is rerun using these revised estimates, and a new projection X_S of energy supply sector outputs derived. And so on.

The sequence of Steps 7-9 described above are repeated until a point is reached where an examination of the output estimates derived on two or more successive rounds suggests that the combined model's operation has converged on a reasonably stable solution.

The current BNL version of the multiregional input-output submodel operates for the 9 United States' census regions, involves 7 energy supply, 8 energy product and 12 non-energy sectors, and employs a Chenery-Moses type of assumption when deriving interregional trade coefficients (see Goettle et al, 1977). However, Isard and Anselin (1980) propose an extension to cover 50 states and a more disaggregated set of non-energy sectors.

Although the above discussion focusses on a specific example of the linkage of interregional input-output and linear programming modules, the same basic procedures would be involved regardless of which sector the linear programming module deals with. That is:

(i) each region's economy is divided into three sets of sectors: (a) non-optimizing sectors N, whose production/trade patterns are determined primarily via the operation of the input-output submodel; (b) optimizing sectors S, whose production/trade patterns are determined primarily via the operation of the linear programming submodel; and (c) dummy sectors P, whose role is to convert the corresponding optimizing sector outputs into a form required to satisfy intermediate and final demands from non-optimizing sectors N and other optimizing sectors S; and

(ii) an iterative solution procedure of the type described on pp. 123-126 is employed.

In addition, the operation of the linear programming submodel can be extended to include more than one sector of the economy. When this is done, it may be necessary or desirable to operate a different programming submodel for each optimizing sector. For example, the agricultural and energy sectors may be judged to be best handled via the use of a programming approach. The most reasonable objective function for the energy sectors may involve minimization of total costs involved in meeting a given set of energy final demands; while for the agricultural sectors, the most reasonable objective function may involve maximization of the total income received by agricultural producers given a set of markets for their outputs. For further discussion in the Australian context, see Appendix 4A.

4.4 Integration of Comparative Cost-Industrial Complex, Programming and Input-Output Models

4.4.1 Results of Comparative Cost-Industrial Complex Analysis as Constraints in Programming Models

The integrated models discussed on pp. 107-121 relating to the entire multiregional economy suffer from two major problems:

(i) the use of a programming model (with interregional input-output relations incorporated as one set of constraints) enables the analyst to identify the regional and/or sectoral distribution of production activities which is optimal given a particular objective function and set of socio-political constraints. However, generally speaking, these results are normative rather than descriptive and may suggest regional distributions for some activities (sectors) which are inconsistent with the operation of basic Weberian-type location principles.

(ii) the models are strictly linear. That is, each assumes that the input-output relations characterizing any given sector are independent of both (a) the scale of its own operation in a given region (i.e., no scale economies/diseconomies are allowed) and (b) the scale of operation in the other sectors in that same region (i.e., no spatial juxtaposition economies/diseconomies are allowed).

Each of these problems can be meaningfully ameliorated through the introduction of an additional set of equality constraints -- equality constraints that reflect the results of comparative cost and/or industrial complex studies conducted for cost-sensitive industries. The steps involved in the resulting integrated framework are depicted in Figure 4.4. They are:

Step 1: *Identification of Sectoral Structure*

The first step involves the identification of those cost-sensitive industries (say $i = 1, ..., k$) whose location patterns are most meaningfully handled via the use of comparative cost and/or industrial complex approaches.

Step 2: *Derivation of Output Projections for Cost-Sensitive Industries*

The next step involves the use of multiregional variants of the comparative cost and/or industrial complex approaches to derive for any given point in time t a projection of the regional distribution of output (and hence capacity expansions) for each of the cost-sensitive industries identified in Step 1.[40]

Step 3: *Identification of Equality Constraints for Cost-Sensitive Industries*

Next, the results derived in Step 2 are incorporated into the following set of equality constraints for each region J (J = A, ..., U) and industry i (i = 1, ..., k):

$$^J X_{i_t} = {}^J\pi_{i_t} X_{i_t} \tag{4.4.1}$$

LINEAR PROGRAMMING MODULE

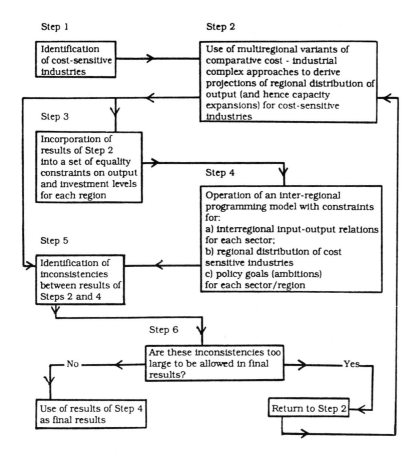

Figure 4.4 Linkage of Comparative Cost-Industrial Complex and Multiregional Programming Analysis

$$^J I_{i_t} = {}^J\Omega_{i_t} I_{i_t} \tag{4.4.2}$$

where $^J X_{i_t}$ and X_{i_t} are the projections of cost-sensitive industry i's output in region J and the nation, respectively, for time period t; $^J I_{i_t}$ and I_{i_t} are the projections of net capacity expansions required in cost-sensitive industry i in region J and the nation, respectively, for the time period t; and $^J \pi_{i_t}$ and $^J \Omega_{i_t}$ are the proportions of X_{i_t} and I_{i_t} respectively allocated to region J.

Step 4: *Derivation of Output Projections for Non-Cost Sensitive Industries*

The next step involves the operation of an interregional programming model of the type discussed on pp. 107-121, making sure that for the point of time t the interregional trade coefficients incorporated in the input-output relation constraints given by (4.3.1) or (4.3.3) are consistent with the regional distribution of cost-sensitive industries given by equation (4.4.1). This yields as one set of results for time period t the regional distribution of national output in each of the remaining non-cost-sensitive industries (i = k+1, ..., n). Where dynamic input-output relations are included in the programming model, then it also yields for each time period t the regional distribution of national capacity expansions in each of the non-cost-sensitive industries.

Step 5: *Identification of Inconsistencies in Output Projections*

Generally speaking, the results derived from an initial run of Steps 1-4 will give rise to inconsistencies. For example:

(i) The regional pattern of cost-sensitive industry outputs incorporated into the set of equations (4.4.1) may be inconsistent with the demands (markets) that the cost-sensitive industries were assumed to provide for each other at the beginning of the comparative cost-industrial complex studies for time period t;[41]

(ii) for any given region, the levels of output of its non-cost-sensitive industries derived for time period t at the end of Step 4 may be inconsistent with the pattern implicit in the set of regional markets estimated for cost-sensitive industry outputs at the beginning of the comparative cost-industrial complex studies for time period t;

(iii) for any given cost-sensitive industry (or complex of such industries) the regional pattern of outputs in other industries (or complexes of industries) derived for time period t in steps 2 and 4 may be inconsistent with the pattern implicit in the set of regional cost differentials estimated when conducting comparative cost-industrial complex analysis for that industry (or complex of industries) for time period t.

Step 6: *Elimination of Major Inconsistencies*

If any of the inconsistencies identified in Step 5 are too large to be permitted in the final results the analyst must use the results derived for other industries in Steps 2 and 4 to reestimate the regional markets (demands) for the output of each cost-sensitive industry. Steps 2-4 are then repeated. That is, new comparative cost-industrial complex studies are conducted to establish new (more logical) levels for cost-sensitive industries in each region for time period t. These results are then employed in a rerun of the interregional programming model, and a new set of results derived for the non-cost-sensitive industries in time period t.

Such reruns may be required for a number of rounds before all major discrepancies are eliminated. No formal proof is available to show that such a rerun procedure will lead to successively smaller and smaller discrepancies, and although such an integrated model was first proposed by Isard et al (1960)[42] there is insufficient experience with operating such a model for any empirical results to be reported here. Nevertheless, the REGAL and MORSE models (see pp. 101-115, and 115-

120, respectively) could be improved considerably if the set of pre-determined (exogenous) share coefficients used to allocate 'non-optimizing' cost-sensitive sectors were to be replaced by a set of endogenous share coefficients derived from the results of comparative cost and/or industrial complex studies.[43]

4.4.2 Linkage of Comparative Cost-Industrial Complex, Input-Output and Programming Models

The integrated model discussed on pp. 121-127 which links an interregional input-output submodel with an individual sector oriented programming submodel also suffers from an inability to handle non-linear inter-dependence between industries (sectors).[44] As a result, it too can be usefully supplemented by the use of comparative cost-industrial complex analyses for cost-sensitive industries (and complexes of such industries).[45] The steps involved in the resulting framework are depicted in Figure 4.5 and represent a combination of (a) those outlined on pp. 77-85 with respect to the linkage of input-output and comparative cost-industrial complex submodels, and (b) those outlined on pp. 121-127 with respect to the linkage of input-output and linear programming submodels. These are:

<u>Step 1:</u> *Identification of Sectoral Structure*

Each region's economy is divided into four sets of sectors: energy supply sectors S, energy product sectors P, cost-sensitive non-energy sectors N' and other non-energy sectors N".

<u>Step 2:</u> *Identification of 'Reasonable' Projections of Relevant National Variables*

The next step involves the identification of exogenous projections of relevant system (national) variables such as: sectoral output, employment, final demand by type, income by type and production.

LINEAR PROGRAMMING MODULE

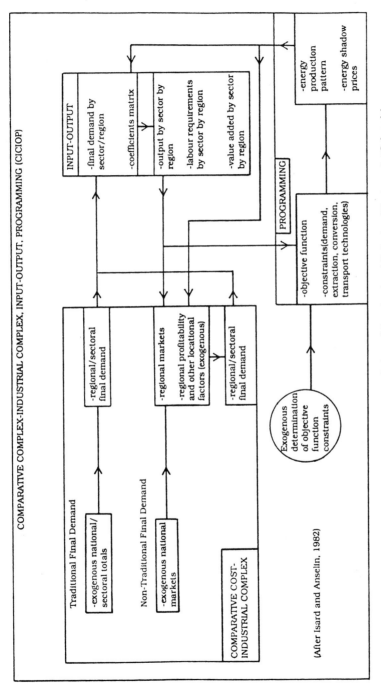

Figure 4.5 Linkage of Comparative Cost - Industrial Complex, Input-Output and Programming Submodels

Step 3: *Derivation of Initial Traditional Final Demand Projections*

The next step involves the disaggregation by region of the national (system) final demand estimates identified in Step 2, and adjustment for net foreign imports to yield a projection of the level of 'traditional' final demand by type by sector by region.

Step 4: *Construction of Appropriately Partitioned Interregional Æ Matrix*

The cost-sensitive non-energy sectors are removed from the interindustry section of the traditional input-output transactions table, and placed within a set of 'non-traditional' final demand columns/primary input rows. The modified interregional technical-trade coefficients matrix Æ can then be partitioned in the manner given by (4.3.16) where the N's are replaced by N'''s.

Step 5: *Derivation of Initial Market Projections for Cost-Sensitive Industries*

The next step involves the estimation of the level of regional demands (intermediate and final) for the output(s) of each of the industries (or parts of industries) identified in Step 4.

Step 6: *Derivation of Initial Non-Traditional Final Demand Projections*

The next step involves the use of the multiregional variants of the comparative cost and/or industrial complex approaches to derive estimates for each region of the level of each industry (or part of industry) placed within the non-traditional final demand column(s) of the input-output submodel.[46]

Step 7: *Consolidation of Traditional and Non-Traditional Final Demand Projections*

Now that the 'traditional' and 'non-traditional' final demand by sector by region have been estimated (in Steps 3 and 6 respectively), they can be consolidated into the set of Y_N, Y_P and Y_S components required for operation of input-output and programming submodels.

Step 8: *Derivation of Initial Output Projections for Non-Energy and Energy Product Sectors*

A first approximation to the output requirements for the non-energy sectors N and energy product sectors P can then be calculated respectively as:

$$X_N'' = (I - \mathcal{Æ}_{N'N'}'')^{-1} Y_N'' \tag{4.4.3}$$

and

$$X_P = \mathcal{Æ}_{PN}'' X_N'' + Y_P \tag{4.4.4}$$

Step 9: *Projection to Functional Energy Service Demand Levels*

The output estimates X_P derived in Step 7 are used to derive estimates of regional functional energy service demand levels ${}^J D_d$ (d = 1, ..., 17) required for operation of the linear programming submodel.

Step 10: *Derivation of Initial Output Projections for Energy Supply Sectors*

The linear programming submodel is run to yield a first approximation of the output X_S of the energy supply sectors and the $\mathcal{Æ}_{SS}$ and $\mathcal{Æ}_{SP}$ submatricies.

Step 11: *Successive Reruns of Input-Output and Programming Modules*

The results of Step 10 are used to derive revised estimates of X_N, X_p and ^JD_d (J = A, ..., U; d = 1, ..., 17). The linear programming submodel is rerun, and new projections of X_S derived. And so on, with Steps 8-10 repeated until a point is reached where the combined operation of the input-output and linear programming submodels converges on a stable set of outputs for the energy product, energy supply and other non-energy sectors.[47]

Step 12: *Identification of Inconsistencies in Output Projections*

The set of output vectors X_N', X_N'', X_p, and X_S derived from a first run through of Steps 1-11 can generally be expected to contain a number of inconsistencies. For example:[48]

(i) for each cost-sensitive non-energy sector, the regional pattern of outputs derived at the end of Steps 6 and 11 may be inconsistent with both (a) the set of markets initially estimated (or assumed) in Step 5, and (b) the set of regional cost differentials calculated in Step 6. This last set of inconsistencies can be expected to be quite marked for energy cost differentials in early rounds.

(ii) for each region, the output of cost-sensitive non-energy sectors derived at the end of Step 6 may be inconsistent with the set of requirements for those outputs implicit in the output vectors X_N'' and X_S derived at the end of Step 11 from the joint operation of the input-output and programming submodels.

(iii) for each region the output vectors X_N' (derived at the end of Step 6) and X_N'', X_p, X_S (derived at the end of Step 11) may be inconsistent with the level of traditional final demand by type of sector estimated for that region at the end of Step 2.

Step 13: *Elimination of Major Inconsistencies*

If any of the inconsistencies identified in Step 12 are judged to be too large to be permitted in the final projections, the analyst must re-estimate one or more of:

(i) the level of traditional final demand by sector by region,

(ii) the pattern of regional markets (demands) for the output of cost-sensitive non-energy sectors,

(iii) the regional cost differentials associated with the operation of one or more cost-sensitive non-energy sectors.

Steps 6-12 are then repeated, and a new set of output estimates derived for each region/sector. These results are checked for inconsistencies. And so on. Once again, a number of such reruns would normally be required before all major discrepancies are eliminated, and no formal proof of convergence has yet been provided.

Although the above discussion relates to the CICIOP model proposed by Isard and Anselin (1980) in which the energy sectors form the basis for the interregional programming submodel, the same set of procedures would be involved were this submodel to focus on another sector. Also, where one or more programming submodels are employed (each focussing on a different sector, or set of interrelated sectors), a similar set of procedures would be required for linking the different submodels.

4.5   Concluding Remarks

This chapter has dealt with the linear programming module and its linkage to the input-output and comparative cost-industrial complex modules of an integrated multiregion model. In the Australian context a linear programming module would be most useful when applied to agricultural sectors in those regions where alternative land uses, crop rotations, etc. are possible. Two linear programming modules have already been developed for Australian agriculture -- one at the Bureau of Agricultural Economics (Longmire, et al, 1979) and the other by a group of scholars at University of New England (Monypenny, 1975, Walker and Dillion, 1976 and Wicks and Dillion, 1978). Although these models would have to be altered somewhat in order to facilitate the integration process, it is clear that the data and expertise exist to get an appropriate linear programming module operational.

NOTES

[1] Although there has been some experimentation with nonlinear multiregional programming models, they are not reviewed here since algorithms capable of solving such programs for any reasonable level of sectoral/regional disaggregation are not generally available.

[2] In doing so we draw heavily upon Isard et al (1960, pp. 415-474), and Isard (1958). For a similar model, see also Stevens (1958).

[3] Since there are k resources and intermediate products, there can be as many as k dummy activities in each region.

[4] Where $h = r+1, ..., m$ it is assumed that the corresponding intermediate product is shipped.

[5] Thus, $\hat{A} = n + k + (U-1)r$ and there are as many as $U\hat{A}'$ activities in the entire system.

[6] Activities which are responsible for trade in transportation may require only one input (namely transportation).

[7] Thus, for each dummy activity $(\hat{A} = 1, ..., k)$ in region J we have the following relation: $-{}^J a_{h_{\hat{A}}} = {}^J a_{h_{\hat{A}}} = 1$, where $h = 1, ..., k$.

[8] After multiplying both sides by minus one and changing the sign of the inequality.

[9] Note that shadow prices can be derived from the dual of the above linear program. These shadow prices show (i) for each resource in a given region, how much the objective function Z would increase as a result of a unit increase in its endowment in that region; and (ii) for each finished product in a given region, how much the objective function Z would increase as a result of a unit decrease in its fixed consumption requirement in that region. See Isard et al (1960, p. 472-474) for the mathematical formulation of this dual program.

[10] This capacity was limited by two factors, namely the supply of underground/surface coal available in that region and the infrastructure available for conducting the given type of mining operation.

[11] Note that this type of formulation assumes that the demand for coal can be specified for each region independently of its delivered price. For some commodities and/or regions such an assumption may substantially reduce the applicability of the model.

[12] Typical resources considered include underground and stripmined coals, crude and shale oils, unprocessed natural gas, uranium ore and

hydropower resources. For a complete list, see Goettle et al (1977) or Isard and Anselin (1980).

[13]Typical intermediate energy forms considered include refined petroleum products, high and low BTU gases, cleaned coal, nuclear fuel and electricity. For a complete list, see Goettle et al (1977) or Isard and Anselin (1980). Some resources are also intermediate energy forms, and when this is the case the corresponding conversion activity is a dummy activity.

[14]Typical functional energy services (i.e., end-use demand categories) considered include ore reduction feedstocks, process heat, electric drive, petrochemicals, automobiles, and space heat, water heat and cooking for both residential and commercial uses. See Goettle et al (1977) or Isard and Anselin (1980) for a complete list.

[15]For other examples and further discussion, see Moses (1960), Nijkamp and Paelinck (1973), Hashim and Mathur (1975), and Mash (1976). In order to reduce the scale of the model still further, the interregional distribution of some sectors may be determined via the use of a comparative cost-industrial complex approach. When this is done, we have an operational version of the integrated approach designated Channel IV in Isard et al (1960). See the discussion on pp. 128-132.

[16]The theory of dynamic input-output analysis was first developed by Leontief (1953) and further refined by Dorfman et al (1958). Miernyk et al (1970) provides the best example of a regional application.

[17]For a formulation of a dynamic interregional input-output model which permits unused capacity, see Isard et al (1960), pp. 716-717 and Isard et al (1969), pp. 989-991.

[18]See Snickars and Granholm (1981) and Lundqvist (1981) for a formulation which permits captital stock to depreciate over time.

[19]See pp. 17-26 for further discussion of the relation between these coefficients.

[20]These include 2 primary product sectors, 2 non-metallic manufacturing sectors, and 2 service sectors which serve primarily national rather than regional markets.

[21]This set of constraints thus represents a modification to equation (4.3.4) above.

[22]See also Isard et al (1960, pp. 684-713), and Chapter 9 below.

[23] Those sectors comprise the agriculture, forestry, mining, non-metallic mineral products, and energy and water provision industries.

[24] In addition, there are the obvious non-negativity constraints on all decision variables (e.g., output, employment, consumption, energy use, imports, exports).

[25] Although the $^{JL}c_i$ coefficients employed are the Chenery-Moses type, they are assumed to be of the form $^{JL}c_i = {}^{JL}\theta_i (1 - {}^L\beta_i)$; where $^L\beta_i$ is the proportion of region L's total demand for sector i's output which are derived from local sources, and $^{JL}\theta_i$ is the proportion of region L's imports of sector i's output that are purchased from region L.

[26] These $^{JL}g_i$ coefficients are calculated in a parallel manner to the $^{JL}c_i$ coefficients just discussed.

[27] These exceptions relate to the regional technical coefficients re: household consumption and the trade coefficients relating to foreign imports and exports. For each of these coefficients, national trends are derived from the MSG model, and each region's coefficients are assumed to follow the same time path as their national counterpart.

[28] The base year $^Ja_{ij}$, $^{JL}c_i$, and $^{JL}g_i$ coefficients are derived from a multiregional input-output model developed by Snickars (1981) using efficient information adding principles.

[29] A similar method is employed to derive time paths of regional energy-output and capital-output coefficients required for the energy and capital supply constraints to be discussed below.

[30] This set of constraints thus represent a modification to equation (4.3.4) above.

[31] See also Isard et al (1960, pp. 684-713), and Chapter 9 below.

[32] For example, see Bargur (1969), Mathur (1972), Batten (1981, 1981a) and Batten and Anderson (1981). Of these, the model proposed by Batten is particularly interesting in that it adopts a hierarchical approach. That is, for a given projection period, a national dynamic input-output model is first employed to identify the maximum sustainable growth rate for the economy as a whole, and the corresponding output levels for each sector (this is generally known as the 'turnpike' solution). A multiregional dynamic input-output model is then employed to identify a set of regional development plans consistent with (a) these national magnitudes, and (b) a set of regional development/welfare goals.

[33] Typical energy supply sectors S include coal mining, crude oil extraction, and refined petroleum products. Typical energy product sectors P include ore reduction feedstocks, motive power, space heat, and electric power. Typical non-energy sectors include livestock and agriculture, durable goods manufacturing, public construction and transport services. For a complete list of sectors, see Goettle et al (1977) or Isard and Anselin (1980).

[34] Since the dummy energy product sectors P are defined in such a way as to produce interregionally non-transportable end products, the $Æ_{PS}$ and $Æ_{PN}$ matrices are block diagonal. See Isard and Anselin (1980) for further discussion.

[35] The submatrix $Æ_{SN}$ is zero, since purchases of energy end-products produced by the P sectors replace the purchase of the corresponding energy resources produced by the S sectors.

[36] See Isard and Anselin (1980) for further discussion.

[37] Dantzig (1976) showed that by expressing the output vectors X_S, X_P, and X_N as functions of the relevant LP variables, the demand constraints in the LP submodel could be reformulated in such a way as to embody the interregional input-output relations employed in the I-O submodel. As a result, the integrated model could be solved via a non-iterative solution procedure. However, since the interative solution procedure (a) reveals more clearly the linkage of the LP and I-O submodels, and (b) allows more frequent checks on the reasonableness of the results, the Dantzig approach is not presented here.

[38] These procedures involve what Goettle et al (1977) designates as a set of I-O to LP interface equations, and are required since there is not always one-to-one correspondence between energy product sector outputs and functional energy service requirements.

[39] This involves the use of a set of LP to I-O interface equations which are responsible for converting the LP results for each region's energy resources (r=1,...,9) and intermediate energy forms (e=1,...,10) into the required I-O results for each region's energy supply sectors S. Many cells within both the $Æ_{SS}$ and $Æ_{SP}$ matrices will be zero because of physical, technological and/or economic restrictions on the outputs which can be produced from any given resource. See Goettle et al (1977) for further discussion.

[40] The steps involved in these approaches, and the circumstances under which one approach should be used in preference to the other are outlined on pp. 54-60 and 73-74. This material is not repeated here.

[41] For further discussion of this source of inconsistencies, see pp. 61-62.

[42] Here the constraints summarized by equation (4.4.2) were designated 'locational efficiency' constraints, and the integrated model was designated Channel IV.

[43] Since the MORSE model is run over three time periods ($t=t$, $t+1$, $t+2$), the comparative cost-industrial complex studies would also need to be conducted over three time periods. Since capacity expansions in the non-cost-sensitive industries are made endogenous in dynamic models, greater inconsistencies can be expected to arise in initial runs between the results of comparative cost-industrial complex studies and the results derived from the operation of the programming model. However, after a few re-runs the skillful analyst should be able to force the overall results to converge in a reasonable manner for each point of time t within the projection period.

[44] That is, no scale economies (diseconomies) are allowed within any given industry and no spatial juxtaposition economies (diseconomies) are allowed to characterize relations between industries.

[45] This approach to integration was first suggested by Isard (1960) and sketched in further detail in Isard et al (1960), Chapter 12 where it was designated a Channel V synthesis. The CICIOP model proposed by Isard and Anselin (1980, 1982) represents a refinement of this channel of synthesis.

[46] Steps 2-6 above resemble closely those given as Steps 1-5 on pp. 77-82 to which the reader is referred for a more detailed discussion.

[47] Steps 8-11 above resemble closely those given as Steps 2-9 on pp.122-125 to which the reader is referred for a more detailed discussion.

[48] These inconsistencies resemble those given on pp. 61-62, to which the reader is referred for a more detailed discussion.

APPENDIX 4A:
MULTIREGIONAL LINEAR PROGRAMMING MODELS OF AUSTRALIAN AGRICULTURE

As discussed in Chapter 4, for some sectors the input-output approach is not the best method for projecting future levels of economic activity. For example, a linear programming model is superior for the energy sector in the United States (see Goettle et al, 1977, and Isard and Anselin, 1980). In the case of Australia, a linear programming model would be superior for the major agricultural sectors.[1] This is so since a programming model will allow substitution between alternative uses of a given acreage of land and other farm inputs to produce a number of alternative sets of outputs. In fact, two linear programming models have already been developed for Australian agriculture.[2] With minor modifications either of these models could be fused with an interregional input-output model. Such in essence requires removing the relevant agricultural sectors $\alpha = 1, \ldots, \alpha'$ with level of operation $^J X_\alpha$ in region J, J=A, ...,U from the input-output structural matrix. A set of agricultural end-products, designated $e = 1, \ldots, e'$, with a pool $^J S_e$ for each region J is then specified. The pool $^J S_e$ (in physical units) is available to meet foreign export demand $^J X_e$ (in phyiscal units), intermediate demands from non-agricultural sectors $n = 1,\ldots,n'$ and agricultural sectors themselves, as well as other final demands $\Sigma\, ^J Y'_e$ from all regions J.

Given (a) a consistent set of prices $^J P_e$,[3] (b) the traditional input-output coefficients of the form $^{JL}a_{en}$, $^{JJ}a_{e\alpha}$, $^{JL}a_{n\alpha}$ and $^{JJ}a_{nn}$, and (c) a set of output coefficients $^{JJ}b_{\alpha e}$ showing the output of end-product e in physical units from a unit of agricultural activity a in region J, the combined input-output linear programming problem can then be stated as:

$$\max \sum_J \sum_e {}^J P_e \, {}^J Z_e$$

(4.A.1)

subject to

$$^J Z_e = {}^J S_e - \sum_L \sum_n ({}^{JL}a_{en}/{}^J P_e){}^L X_n - \sum_L ({}^J Y_e/{}^J P_e)$$

$$- \sum_L \sum_\alpha ({}^{JL}a_{e\alpha}/{}^L P_e){}^L X_\alpha \qquad (4.A.2)$$

for all e, J;

$$^J S_e = \sum_\alpha {}^{JJ}b_{\alpha e} {}^J X_\alpha \qquad (4.A.3)$$

for all e, J;

$$\sum_e ({}^J P_e {}^{JJ}b_{\alpha e} {}^J X_\alpha) - \sum_L \sum_h {}^{JL}a_{h\alpha} {}^J X_\alpha \geq 0 \qquad (4.A.4)$$

for all a, J where h = 1,...,e'; n = 1,...,n';[4]

$$X_n = [I - A_{nn}]^{-1} + \sum_\alpha a_{n\alpha} X_\alpha \qquad (4.A.5)$$

as well as resource and other constraints specific to agriculture.[5] For example, we may wish to impose upper bounds on the stock of land, labour capital and livestock available for use in different activities (pasture, cropping, sheep and cattle raising) in each region.[6] And so on.

One way to better understand the nature of this integrated model is to conceive it as proceeding via the following steps:

Step 1. Calculate an initial estimate of the output of each non-agricultural sector as:

$$X_n = [I - A_{nn}]^{-1} Y_n$$
$$\text{(Unx1)} \quad \text{(UnxUn)}$$
(4.A.6)

Step 2. Calculate an initial estimate of the demand for agricultural end-product e as:

$$^J D_e = \sum_L \sum_n (^{JL} a_{en} / ^L P_e) \, ^L X_n + \sum_L (^L Y'_e / ^L P_e) \quad \text{for all e, J.} \quad (4.A.7)$$

Step 3. For the moment, ignore the last term on the RHS of equation (4.A.2) so that taking into account equation (4.A.7) we have:

$$^J Z_e = {}^J S_e - {}^J D_e \qquad \text{for all e, J.} \qquad (4.A.8)$$

Step 4. Using equation (4.A.3), we can next rewrite equation (4.A.8) as:

$$^J Z_e = \sum_\alpha {}^{JJ} b_{\alpha e} \, {}^J X_\alpha - {}^J D_e$$
(4.A.9)

for all e, J.

Step 5. So given $^J D_e$, to max $\sum_J \sum_e {}^J P_e \, {}^J Z_e$ subject to a set of resource constraints is to

$$\max \sum_J \sum_e {}^J P_e \, (\sum_\alpha {}^{JJ} b_{\alpha e} \, {}^J X_\alpha - {}^J D_e)$$
(4.A.10)

subject to the same set of resource constraints. This then yields an initial estimate $^J X_\alpha$ of the level of agricultural activity α ($\alpha = 1,...,\alpha'$; J=A,...,U).

Step 6. But to produce $^J X_\alpha$ requires inputs from both the regular input-output sectors and agricultural sectors. Hence we recalculate the output of non-agricultural sectors as:

$$\underset{(\text{Unzl})}{X'_n} = \underset{(\text{UnxUn})}{\left[I - A_{nn}\right]^{-1}} \underset{(\text{Unx1})}{Y_n} + \underset{(\text{UnxU}\alpha)}{A_{n\alpha}} \underset{(\text{U}\alpha\text{x1})}{X_\alpha} \quad (4.\text{A}.11)$$

where a typical element $a_{n\alpha}$ of the matrix $A_{n\alpha}$ refers to the dollar's worth of input sector n in region J required per unit level of operation of activity α in region L.

Step 7. Next, the demand for agricultural end-product e can be re-estimated as:

$$^J D'_e = \sum_L \sum_n (^{JL}a_{en}/^L P_e)^L X'_n + \sum_L (^L Y_e/^L P_e)$$

$$+ \sum_L \sum_e (^{JL}a_{e\alpha}/^L P_e)^L X_\alpha \quad \text{for all e, J.} \quad (4.\text{A}.12)$$

Step 8. A revised estimate of agricultural sector activity levels $^J X'_\alpha$ can then be derived using steps 3 through 5 above, after substituting $^J D'_e$ for $^J D_e$.

Step 9. In the next run, we substitute $^J X'_\alpha$ for $^J X_\alpha$ in equations (4.A.3), (4.A.10) and (4.A.11), and continue with such re-runs until the costs of further reiterations are not justified by improvement in the so-called 'accuracy' of results.

NOTES

[1] An earlier version of this appendix was included in Isard and C. Smith, 1986.

[2] One of these models has been constructed by the Bureau of Agricultural Economics (see Longmire, et al, 1979), the other by a group of research scholars at the University of New England (see Monnypenny, 1975, Mandeville, 1976, Walker and Dillion, 1976, and Wicks and Dillion, 1978).

[3] These prices are used to convert standard input-output coefficients (in dollar terms) into physical units, and hence represent delivered prices. We take the delivered price $^J P_e$ for farmers in region J to be the price at the port of export for his product. This latter price is equal to (a) the price in the part of the world market to which he exports, less (b) the transport cost incurred in reaching this market. To be consistent with input-output methodology we assume that the farmer incurs a constant transport cost per unit of output.

[4] This constraint ensures non-negative profits for each agricultural activity.

[5] In traditional manner, we impose no constraints on the operation of the regular input-output sectors.

[6] Land types may include unimproved, intermediate, and improved pasture as well as cropping land; labour types may include owner-operator and family labour as well as permanent and casual employees; capital types typically include plant and machinery as well as equipment. Livestock typically comprises sheep of different breeds, ages and sexes, and cattle of different ages and sexes. The current model in the Bureau of Agricultural Economcs includes 10 pasture activities, 16 cropping activities, 23 sheep raising activities and 13 cattle raising activities. It treats 15 agricultural end-products (wool, 2 types of lamb, 2 types of mutton, live sheep, 2 types of beef and 7 types of crops).

CHAPTER 5
A DEMOGRAPHIC MODULE AS A COMPONENT OF AN INTEGRATED MULTIREGIONAL MODEL

5.1 Introduction

This chapter is concerned with the development of a demographic module which is capable of forming a component of an integrated multiregional model. In Section 5.2 an outline is given of three alternative types of multiregional demographic models: (a) simple components of change models, (b) cohort survival models, and (c) microsimulation models. Since the interregional migration component of the demographic module is of most interest to regional scientists, Section 5.3 explores a number of alternative approaches for projecting the pattern of migration flows within a multiregion system. In Section 5.4 an outline is given of the steps involved in integrating multiregional input-output and demographic modules. The operation of a more fully integrated multiregional model comprising: (a) an interregional input-output module, (b) a comparative cost-industrial complex module, (c) an interregional programming module, and (d) a multiregional demographic module is discussed in Section 5.5. Finally, some concluding remarks are made in Section 5.6.

5.2 Outline of Multiregional Demographic Models
5.2.1 Simple Components of Change Approach

Methods of forecasting population of multiregional systems are based on analysis of the following major elements of population change for each region: natural increase (or decrease), and net inmigration (or outmigration). The relationship may be written in the general form:

$$^J\underline{P}_{t+\theta} = {}^J\underline{P}_t + {}^J N_\theta + {}^J M_\theta \qquad (J=A,\ldots,U) \qquad (5.2.1)$$

where $^J P_{t+\theta}$ = population of region J in forecast year t+θ;
$^J P_t$ = population of region J in base year t;
$^J N_\theta$ = net natural increase in region J during period θ; and
$^J M_\theta$ = net migration into region J during period θ.

In crude form, the natural increase component $^J N_\theta$ can be obtained as:

$$^J N_\theta = {}^J b_\theta \, {}^J P_t - {}^J d_\theta \, {}^J P_t \qquad (J=A,\ldots,U) \qquad (5.2.2)$$

where $^J b_\theta$ = the crude birth rate expected in region J during period È; and
$^J d_\theta$ = the crude death rate expected in region J during period θ.

These expected crude birth and death rates may be determined by a subjective analysis of past trend, or via some extrapolation technique.[1] Alternatively, official national projections of these rates may be taken as the base and attempts made to forecast regional variations around these rates -- again on the basis of an examination of past trends.[2] Finally, a more theoretically appealing approach would attempt to relate projections of national (and/or regional) fertility -- and hence crude birth rates -- to prevailing social and economic conditions.[3] For the purposes of this book however, it is assumed that a reasonable projection of such rates can be derived from the appropriate demographic experts.

In crude form, the net migration component $^J M_\theta$ is obtained as:

$$^J M_\theta = \left(\sum_{\substack{J=A \\ J \neq L}}^{U} {}^{LJ} m_\theta \, {}^J P_t \right) + {}^J f_\theta \qquad (J=A,\ldots,U)$$

$$(5.2.3)$$

where $^{LJ} m_\theta$ = the net proportion of the population in region L at time t who have moved to region J by time t+θ;[4] and

$^J f_\theta$ = the net migration into region J from outside the multiregional system during period θ.

The set of equations (5.2.1) - (5.2.3) can be expressed more compactly using matrix notation. That is:

$$\underline{P}_{t+\theta} = [\,B - D + M\,]\,\underline{P}_t + F \qquad (5.2.4)$$

where $\underline{P}_{t+\theta}$ = a (Ux1) column vector of population levels, with a typical element $^J\underline{P}_{t+\theta}$;

$\underline{P}_t$ = a (Ux1) column vector of population levels, with a typical element $^J\underline{P}_t$;

B = a (UxU) matrix with crude birth rates ($^A b_\theta$, ..., $^J b_\theta$, ..., $^U b_\theta$) along the main diagonal, and zeros elsewhere;

D = a (UxU) matrix with crude death rates ($^A d_\theta$, ..., $^J d_\theta$, ..., $^U d_\theta$) along the main diagonal, and zeros elsewhere;

M = a (UxU) matrix of net migration rates, with typical element $^{LJ}m_\theta$; and

F = a (Ux1) column vector with typical element $^J f_\theta$.

The level of national foreign immigration may be projected subjectively on the basis of past trends or recent policy statements made regarding target levels of immigration during the projection period. These national levels may then be allocated among regions via the use of share coefficients derived from base year data,[6] or projected on the basis of past trends.[7] The internal migration rates $^{LJ}m_\theta$ are of most interest in this book however, since it is through them that most effective linkage between economic and demographic modules can be achieved. Two simple approaches which may be employed to project these rates include:

(1) selection of a base period (say the period covered in the latest population census data) and use of the migration rates prevailing in this period; and

(2) identification of past trends in migration rates (eg. by fitting of a trend line or curve to data by use of freehand, graphic, or mathematical methods) and extrapolation of such trends to obtain estimates of future migration ratios.

However, the fundamental postulate of such approaches -- namely that future trends can be identified solely on the basis of past relationships -- is not generally very realistic for the migration component of population change. Migration patterns tend to be highly volatile, responding much more quickly (than other components of population change) to changes in economic, social and political forces affecting the different regions in the system.

As a result, forecasts of future migration patterns (and hence projections of the migration rates underlying such patterns) should be linked in some way to economic variables. Some approaches which allow such linkages are discussed in Section 5.4.

5.2.2 Cohort-Survival Approach

While the simple components of change model may be used to provide crude population projections for a system of regions, a more accurate set of projections can be expected from the use of an approach which recognizes that any given region's overall birth, death and migration rates vary depending on changes in the composition of its population -- particularly its age-sex structure. One such approach is the 'cohort-survival' technique.[8]

To show how this approach works, first let us consider the breakdown of the population in terms of age groups. Let us, for example, disaggregate the overall population in terms of groups spanning θ years. (There will be d such age groups in region J).[9] Denoting the level of population of age group d in region J at time t by $^J\underline{P}_{dt}$, region J's population level at time t can thus be represented by a ($d \times 1$) column vector $^J\underline{P}_t$ with elements ($^J\underline{P}_{1t}, ..., ^J\underline{P}_{dt}, ..., ^J\underline{P}_d$).

DEMOGRAPHIC MODULE

Now let us consider the <u>birth component</u> of change, and denote by $^Jb_{d\theta}$ the birth rate over time span θ of persons in age group d in region J.[10] The total number of births in J during period θ will therefore be $\Sigma\ ^Jb_{d\theta}\ ^JP_{dt}$. Using matrix notation, the set of population level changes in J due to births, designated $\Delta^J\underline{P}_\theta$, is given by:

$$\Delta^J\underline{P}_\theta = {}^JB_\theta \cdot {}^J\underline{P}_t \qquad (5.3.1)$$

where B_θ = a (dxd) matrix, designated the 'birth operator' with non-zero entries ($^Jb_{1\theta}, ..., {}^Jb_{d\theta}, ..., {}^Jb_{d\theta}$) in the first row only.[11]

Extending to U regions (J=A, ..., U) we have:

$$\Delta\underline{P}_\theta = B_\theta \cdot \underline{P}_t \qquad (5.3.2)$$

where $\Delta\underline{P}_\theta$ = a (Udx1) column vector which can be partitioned into U column vectors, each of the form given by $\Delta^J\underline{P}_\theta$;

B_θ = a (UdxUd) block diagonal matrix with a typical diagonal submatrix of the form given by $^JB_\theta$; and

$\underline{P}_t$ = a (Udx1) column vector which can be partitioned into U column vectors, each of the form given by $^J\underline{P}_t$.

The next step is to consider the <u>survivorship component</u> of change. Here we denote by $^Js_{d\theta}$ the proportion of region J's population in age group d who survive to the next age group, namely d+1, during a time span of length θ.[12] The number of survivors from group d in region J is thus equal to $(^Js_{d\theta}) \cdot (^JP_{dt})$. Excluding the effects of births and migration this is, in general, also equal to the population level of the next age group, namely d+1, at time t+θ. However, there are two exceptions: (a) in the first age group this level will, by definition, be equal to zero; and (b) in the last age group d, this level will include survivors in that age group itself $(^Js_{d\theta}) \cdot (^J\underline{P}_{dt})$ as well as those from the penultimate age group d-1.

Using matrix notation, the vector of new population levels in region J due to the effects of aging $^J\underline{P}'_{t+\theta}$ is given by:

$$^J\underline{P}'_{t+\theta} = {^JS_\theta} \, {^J\underline{P}_t} \qquad (5.3.3)$$

where $^JS_\theta$ is a $(d{\times}d)$ matrix, designated the survivorship operator, with all entries equal to zero except those (a) immediately below the main diagonal, or (b) in the lower right corner cell. These non-zero entries correspond to the survivorship ratios ($^Js_{1_\theta}$, ..., $^Js_{d_\theta}$, ..., $^Js_{d\theta}$) defined earlier.[13] Extending to U regions (J=A, ..., U) we have:

$$\underline{P}'_{t+\theta} = S_\theta \cdot \underline{P}_t \qquad (5.3.4)$$

where S_θ = a $(Ud{\times}Ud)$ block diagonal matrix with a typical diagonal submatrix of the form given by $^JS_\theta$; and

$\underline{P}_{t+\theta}$ = a $(Ud{\times}1)$ column vector which can be partitioned into U column vectors, each of the form given by $^J\underline{P}'_{t+\theta}$.

Finally, we consider the <u>migration component</u> of change and denote by $^{LJ}m_{d_\theta}$ the proportion of population in age group d in region L who (on net) migrate to region J during the period θ (and are thus found in age group d+1 at time t+θ).[14] The net number of migrants in group d arriving in region J is thus equal to

$$\sum_{L=A}^{U} (^{JL}m_{d\theta})(^L\underline{P}_{dt}) + {^Lf_{d\theta}} \qquad (L{\neq}J)$$

$$(5.3.5)$$

where $^Lf_{d_\theta}$ denotes the level of net foreign immigration into region L by persons in age group d. Excluding the effects of natural increase, the first term in this expression is, in general, also equal to the population change in the next age group d+1 that can be attributed to migration.

However, there are two exceptions: (a) in the first age group, this change will (by definition) be equal to zero; and (b) in the last age group d, we need to include net internal migrants who at time t were in that age group itself $(^{LJ}m_{d\theta})(^{L}P_{dt})$ as well as those who were in the penultimate age group d-1.

Using matrix notation, the (dx1) vector of net population changes in region J due to migration flow from L is given by $(^{LJ}M_\theta)(^{L}\underline{P}_t)$ -- where $^{LJ}M_\theta$ is a (dxd) matrix of age-specific migration rates, with all entries equal to zero except those (a) immediately below the main diagonal, or (b) in the lower right corner cell. These non-zero entries correspond to the net migration rates $^{LJ}m_{d\theta}$ (d=1, ..., d) defined earlier.[15] Extending to U regions (J=A, ..., U), we have

$$\Delta \underline{P}'_\theta = M_\theta \cdot \underline{P}_t + F_\theta \qquad (5.3.6)$$

where $\Delta \underline{P}'_\theta$ = a (Udx1) column vector with typical element $\Delta \underline{P}'_{d\theta}$ showing the change in region J's population due to migration in age class d over the period θ. This vector can be partitioned into U column vectors (each of order dx1), a typical subvector being $\Delta \underline{P}'_\theta$;

M_θ = a (UdxUd) matrix which can be partitioned into (UxU) submatrices -- each of order dxd. The typical off-diagonal submatrix is of the form given by $^{LJ}M_\theta$, while all the diagonal submatrices are null matrices; and

F_θ = a (Udx1) column vector of net foreign immigration levels which can be partitioned into U column vectors, each of order dx1 -- a typical subvector being $^J F_\theta$ with elements ($^J f_{1\theta}$, ..., $^J f_{d\theta}$, ..., $^J f_{d\theta}$).

The effects of births, deaths and migration may be combined simply by adding the matrices B_θ, S_θ, and M_θ to derive an overall 'growth operator' G_θ. That is, we have:

$$\underline{P}_{t+\theta} = G_\theta \cdot \underline{P}_t + F_\theta \qquad (5.3.7)$$

where G_θ is a $(UdxUd)$ matrix which can be partitioned into (UxU) submatrices. The typical off-diagonal submatrix is of the form $^{LJ}M_\theta$; while the typical diagonal submatrix is of the form derived from the addition of $^JB_\theta$ and $^JS_\theta$.[16]

This completes the basic cohort-survival projection technique. However, it can be extended to incorporate further disaggregation of the population beyond age groups. For example, the population can be broken down in terms of sex, occupational group, ethnic background, marital status, employment status, income group, etc. Each of such potential characterisations basically involves augmenting the size of the matrices involved in determining the basic growth operator G_θ.[17]

However, regardless of the level of disaggregation, the results obtained from cohort-survival techniques can be expected to be only as accurate as the birth, death and migration rates assumed over the forecast period.[18] Precision of calculation and further disaggregation cannot compensate for limitations in the appropriateness of these assumed rates.

Once again, national projections of fertility and mortality trends can be taken as given -- this time for each demographic class (age-sex group, or age-sex-marital status group, etc) -- and regional differences from these projections determined by a subjective analysis of past trends, or some extrapolation technique.[19] Also, we may take as given the national level of net foreign immigration. However, for use in cohort-survival models this level must be disaggregated in terms of its demographic class composition. This may be done, for example, by assuming no change in this composition from some base year period (Olsen, et al, 1977) or by use of a set of share coefficients derived from an examination of past foreign immigration flows (Shryock, Siegel, et al, 1976). Once this is done, the regional allocation of these migrants can be handled in a manner comparable to that discussed on page 151.

However, once again the most critical area of concern to regional scientists is the method employed for projecting the demographic class-specific internal migration rates $^{JL}m_{d\theta}$ $(d=1,...,d;L,J=A,...,U)$. Simple approaches such as the use of constant rates derived from base year data, or extrapolation on the basis of past trends may be used. However, such approaches are not of major interest in this book because they do not recognize the responsiveness of these migration rates to economic, social and/or political conditions in the different regions within the system,[20] and hence do not permit development of a complete set of linkages between the economic and demographic components of an integrated modelling effort. Some approaches which may be employed to incorporate such responsiveness into migration rate (or level) projections are discussed on pp. 180-185.

5.2.3 Microsimulation Approach [21]

The two approaches to population projection discussed above deal with aggregate data. Even when a cohort survival model disaggregates population into different demographic classes, it still operates with regional aggregates. The microsimulation approach, however, deals with micro-level (ie. decision unit) data. That is, it starts with a sample representation of each region's population. The sample population is divided into its fundamental microcomponents or decision-making units (such as individuals and households). Each of these microcomponents is characterised by a set of attributes (age, sex, marital status, number of offsprings, etc. for individuals, and size, income by source, etc. for households). Further, these microcomponents are treated as embedded in environments that condition their behaviour. These environments comprise both relationships to other microcomponents (eg. an individual's relationship to his household) and characteristics of each region's economic, social and political performance.

Next, attributes of each sample microcomponent are 'aged through time' using <u>behavioural modules</u> that take into account both attributes of the microcomponent in the sample period and changes in the attributes of the conditioning environment during the projection period. For example, the demographic component of the DYNASIM model developed by Orcutt <u>et al</u> (1976) includes behavioural modules:

(1) for incrementation of each individual's age and educational attainment;

(2) for computing the probability that each individual will change his household status (if a person is selected to leave, then the attributes of the corresponding households are modified accordingly);

(3) for computing the probability that each married couple will get a divorce (if a couple is selected for divorce, then each moves into a separate household and any dependent children allocated among these new households according to a set of empirically derived rules);

(4) for computing the probability that each unmarried individual will change his/her marital status (if a person is selected for marriage, then she enters the marriage union sector where she is matched with a partner. Each partner's household status is then changed accordingly);

(5) for computing the probability that each individual will die (if a person is selected to die, then she is excluded from subsequent calculations and the marital status of her spouse, if one exists, is changed accordingly); and

(6) for computing the probability that a household moves from its initial region of residence (if a household is selected to move, then the new region of residence is also determined probabilistically).

Since only the simplest, least realistic, microcomponent based models can be solved by analytic or transition matrix methods, simulation is the cost efficient method of solution.[22] Further, since many of the required behavioural modules are stochastic (ie. the probability of occurence of certain attributes, rather than the occurences themselves, are generated) a Monte Carlo type of simulation is typically required.

Once the sample has been 'aged' through time in this manner, these projection period results are then 'blown-up' to the required macro-magnitudes for each region using appropriate scaling factors.

The performance of a micro-simulation based model depends on three main factors: (a) the accuracy with which the initial sample represents each region's population; (b) the accuracy with which the behavioural modules project changes in individual and household attributes; and (c) the degree to which the chance Monte Carlo variation in the micromodel reflects chance variation in the real world. (Caldwell and Saltzman, 1981). Assuming that the appropriate experts can adequately control for the first and third of these sources of error, and that appropriate behavioural modules can be developed for simulating demographic processes other than interregional migration, we now turn to a discussion of the means available for deriving the probabilities required for operation of the migration submodule.

One could derive the probability that a given household (or individual) migrates from one region to another on the basis of past microdata (either by assuming such probabilities remain constant for a selected base year sample, or by allowing such probabilities to change taking into account trends apparent in past samples.)[23] A more behaviourally appealing approach would relate the probabilities derived from past microdata to past macrodata on the economic conditions in the regions of origin and termination, as well as to the given household's (or individual's) attributes.[24] Then by employing one of the migration models to be discussed in Section 5.3, such probabilities could be projected into the future on the basis of forecasts of regional economic conditions derived from other modules comprising the integrated model. Unfortunately, however, implementation of this type of approach requires an adequate time series of micro-based migration data. Where such is not available, then perhaps the benefits claimed for the microsimulation approach at the national level are lost when employed for multiregional population projection.[25] In any case, implementation of this approach requires that a public use sample be made available from at least one national population census (or the equivalent) -- and that such a sample be drawn in such a way that it adequately represents the population of each region as well as the nation. Such a public use sample has not yet been released in Australia, so this approach cannot be adopted for this country.

5.3 Alternative Approaches to Modelling Migration[26]
5.3.1 Gravity Model Approach
a. Unconstrained Model for Aggregate Migration

As its name implies, the gravity model approach was originally developed as an application of Newtonian physics. One relation that the latter states is that the potential energy E generated by two bodies of masses M_1 and M_2, separated by a distance d'', will be equal to

$$E = g\, M_1\, M_2 / d'' \qquad (5.3.8)$$

where g is a gravitational constant.
Furthermore, if we square the distance d'' in the denominator of (5.3.8) we have

$$F = g\, M_1\, M_2 / (d'')^2 \qquad (5.3.9)$$

where F is the familiar gravitational force that each mass exerts on the other.

In parallel fashion, the gravity concept of spatial interaction[27] states in its simple form that the interaction ^{JL}M between two regions J and L will be directly proportional to the "masses" ^{J}Q and ^{L}Q of these regions (eg. population, employment demand, output levels, etc.) and inversely proportional to some function f of the 'distance' $^{JL}d''$ between them.[28] Using the subscript θ to refer to the time period during which the interaction takes place, we have:

$$^{JL}M_\theta = g\, ^{J}Q_\theta\, ^{L}Q_\theta\, f(^{JL}d'') \qquad (5.3.10)$$

Several points need to be clarified with respect to the general formulation (5.3.10), particularly when the interaction involved is net interregional migration:

(1) Firstly, we need to decide upon variables which should be included as the mass variables $^{J}Q_\theta$ and $^{L}Q_\theta$. Some useful candidates include employment opportunities, per capita income levels and gross domestic product.[29] An important question here is whether $^{J}Q_\theta$ should be a different variable (or set of variables) than $^{L}Q_\theta$ -- since $^{J}Q_\theta$ should be a measure of the capacity (or propensity) for interaction generation by region J; while $^{L}Q_\theta$ should be a measure of the capacity (or propensity) for interaction absorption by region L.

To distinguish the 'push-out' factors underlying $^{J}Q_\theta$ from the 'pull-in' factors underlying $^{L}Q_\theta$, these masses can be replaced by $^{J}O_\theta$ and $^{L}D_\theta$, respectively. Thus, we derive:

$$^{JL}M_\theta = g\, ^{J}O_\theta\, ^{L}D_\theta\, f(^{JL}d'') \qquad (5.3.11)$$

Variables which might be included in the origin region 'mass' $^{J}O_\theta$ measure then include labour surplus or unemployment levels, population density, crime levels, and growth rates in per capita income.[30] On the other hand, the destination region 'mass' $^{L}D_\theta$ might include variables relating to labour deficit, growth of employment opportunities, wage rates, industrial structure, and climate.[31,32]

(2) Secondly, a related question is what exponent should be placed on the $^{J}O_\theta$ and $^{L}D_\theta$ measures selected. In equation (5.3.11), these exponents are set at unity. However, more generally each of these 'mass' variables may be raised to a power other than unity. Thus, we derive:

$$^{JL}M_\theta = g\,(^{J}O_\theta)^\alpha\,(^{L}D_\theta)^\beta\,f(^{JL}d'') \tag{5.3.12}$$

where the α and β exponents may be interpreted as measures of the responsiveness of the migration flow $^{JL}M_\theta$ to the 'mass' measures $^{J}O_\theta$ and $^{L}D_\theta$, respectively.[33]

(3) Thirdly, the problem of timing arises. Since $^{JL}M_\theta$ refers to a migration flow during period θ (that is, from t to t+θ) and the mass variables $^{J}O_\theta$ and $^{L}D_\theta$ are (generally speaking) not constant over time, a decision needs to be made concerning the point(s) of time the variables chosen to represent these masses should refer. For example, should migration flows between time t and t+θ respond to origin region characteristics (summarized by $^{J}O_\theta$) at the end of this period (ie. t+θ) or at the beginning of this period (ie. t) or at an intermediate point within this period,[34] and so on. One simple way to resolving problems of this sort is to take $^{J}O_\theta$ (or $^{L}D_\theta$) to be a simple average of the selected origin (or destination) region characteristics over the period t to t+θ.

(4) Fourthly, the issue arises of the appropriate form for the distance function $f(^{JL}d'')$ in equations (5.3.11) and (5.3.12). We have already noted that the use of simply $(^{JL}d'')^{-1}$ suggests a potential energy type of relationship between the two masses $^{J}O_\theta$ and $^{L}D_\theta$; while the use of $(^{JL}d'')^{-2}$ suggests a gravitational-type force exerted by one mass on the other. However, we may use the more general function

$$f_1(^{JL}d'') = (^{JL}d'')^{-\gamma} \tag{5.3.13}$$

and let observations on past interregional migration flows suggest the most appropriate value for the exponent γ.[35] But then, there is no theoretical justification for the use of the power function (5.3.13) rather than a negative exponential function:[36]

$$f_2\ (^{JL}d") = e\ ^{-\beta(^{JL}d")} \tag{5.3.14}$$

Once again, it may be argued that we may let the choice between (5.3.13) and (5.3.14) be determined during the process of calibration -- by selecting that function, together with its associated parameters, which produces the 'best' fit between the output of the model and the observed values.[37]

(5) Fifthly, the issue arises of the appropriate measure of distance d". The most commonly used measure is physical distance, although others (such as combined relocation-travel-job search costs) may be used where the required data is available. However, since each region's population is not concentrated at a single point nor distributed uniformly within its boundaries, we need to select a reference point within each region from which to measure interregional distances.

Given the high level of metropolitan primacy within most Australian states, a logical set of reference points would be the state capitals -- especially since each lies fairly close to its more theoretically appealing population centroid.[38]

(6) Finally, and perhaps most importantly, we come to the issue of how to forecast the mass variables $^{J}O_\Omega$ and $^{L}D_\Omega$ for the projection period Ω. If a single variable was selected to represent one or both of these masses, then we could derive the required forecasts via extrapolation on the basis of past trends for this variable in each region. However, when the selected variable is one that can be generated (directly or indirectly) from operation of an input-output module, then we can reasonably expect that the use of an integrated model including

both a demographic and an input-output module would produce superior projections of interregional migration. Similarly, when the selected variable is one that (for the most part) results from the operation of the labour market, then we may expect that the use of an integrated model including an econometric module would produce superior projections of interregional migration. See pp. 180-185 and 238-240 for further discussion. Regardless of the degree of accuracy with which projections are able to be made for the selected variable, one is ignoring the influence of all other variables on migration patterns. Hence, the degree of accuracy obtained in migration forecasts will depend on the 'correctness' of the particular variable selected.[39] Alternatively, if a number of different variables were selected to represent the regional masses then we in effect have:

$$^J O_\Omega = \sum_{i=1}^{n} {}^J w_{i\Omega} \, {}^J X_{i\Omega}$$

and (5.3.15)

$$^L D_\Omega = \sum_{j=1}^{m} {}^L v_{j\Omega} \, {}^L Y_{j\Omega}$$

where ${}^J w_{i\Omega}$ = the weight attached to the 'push-out' factor ${}^J X_{i\Omega}$ associated with region J; and

${}^L v_{j\Omega}$ = the weight attached to the 'pull-in' factor ${}^L Y_{j\Omega}$ associated with region L.[42]

In order to forecast the mass variables ${}^J O_\Omega$ and ${}^L D_\Omega$ we now need to forecast the set of 'push-out' factors ${}^J X_{i\Omega}$ (i=1,...,n) and 'pull-in' factors ${}^L Y_{j\Omega}$ (j=1,...,m) as well as their respective weights.

Some (or all) of the ${}^J X_{i\Omega}$ and ${}^L Y_{j\Omega}$ variables may be able to be generated via the use of either an input-output module, or an econometric module or both -- hence the above comments on the usefulness of an integrated model are also appropriate here. However, the derivation of the weights $^Jw_{i\Omega}$ and $^Lv_{j\Omega}$ is a more difficult task -- especially if we wish to allow such weights to vary from the base period weights $^Jw_{i\theta}$ and $^Lv_{j\theta}$, respectively. See pp. 174-180 for one means via which this problem could potentially be resolved -- given sufficient data to run an econometric model of regional in-and out-migration flows.[41]

b. Unconstrained Model for Migration by Demographic Class

The gravity model just described can be used to forecast net interregional migration flows for the population as a whole. However the analyst may be interested in projecting such flows for each of a number of demographic classes d (d=1, ..., d), or he may believe that a more disaggregated approach is required given the differences between demographic classes in terms of:

(1) the origin-region characteristics considered important when making out-migration decisions and/or the weights attached to these characteristics;

(2) the destination-region characteristics considered important when making in-migration decisions and/or the weights attached to these characteristics; and

(3) the importance of distance as an influence in their migration decisions.

These differences will be reflected in the $^JO_\theta$, $^LD_\theta$ and $f(^{JL}d'')$ terms in equation (5.3.12) and suggest the use of a different gravity model for each demographic class d. That is, that we construct d gravity models of the form:[42]

$$^{JL}M_{d\theta} = g_d \, (^JO_{d\theta})^\gamma \, (^LD_{d\theta})^\delta \, f_d(^{JL}d'') \qquad (5.3.16)$$

$LD_{d\theta}$ = a measure of the capacity (or propensity) of region L to absorb inmigrants in demographic class d during period θ.

Once again, there are a number of problems (issues) that need to be resolved before this set of models can actually be implemented:

(1) Firstly, the mass variables $JO_{d\theta}$ should now include those 'push-out' factors in region J considered relevant by members of demographic class d; while the mass variable $LD_{d\theta}$ should now include those 'pull-in' factors in region L considered relevant by members of demographic class d. Thus, we may have some factors included in the $JO_{d\theta}$, $LD_{d\theta}$ for persons in one demographic class (ie. for d=d') and not for another (ie. for d=d"). Further, when a particular factor -- say growth in employment opportunities -- is selected for in inclusion in $JO_{d\theta}$ and $LD_{d\theta}$, we have the problem of identifying which employment opportunities are considered relevant by members of demographic class d. That is, we have to ask ourselves whether people

(a) just consider the 'benefits' from migration accruing to members of their present demographic class, or

(b) take into account the possibility of changes in their demographic class affiliation over their lifetime -- either because of natural progression from one demographic class to another (eg. movement from one age group to another), or because of expected progressions (eg. movement from one occupational group to another as skill levels increase over time).

The latter 'lifetime' calculations are suggested by the 'human capital' theory of migration, and can be introduced into the model outlined above by defining for each demographic class d a set of classes D which could reasonably be expected to be reached from d (and which are not so 'distant' as to be eliminated from consideration once a reasonable discount factor is employed). The general measure of labour market conditions selected as an origin (or destination) region mass variable, say employment opportunities $JE_{d\theta}$ in region J, would then be taken to conditions selected as an origin (or destination) region mass variable, say employment opportunities $^JE_{d\theta}$ in region J, would then be taken to be a weighted average of employment opportunities $\Sigma\ h_d\ ^JE_{d\theta}$ for members of all demographic classes d in D, rather than just employment opportunities for members of demographic class d. Once again, supplementary modes of analysis are required to derive the set of weights h_d to be used when constructing the average -- and also in identifying appropriate d, D pairs.

(2) Secondly, we need to project $^JO_{d_\Omega}$ (d=1,...,d) and $^LD_{d_\Omega}$ (d=1,...,d) rather than simply $^JO_\Omega = d\Sigma\ ^JO_{d\Omega}$ and $^LD_\Omega = d\Sigma\ ^LD_{d\Omega}$. Again, many of the demographic class specific variables (eg. employment opportunities, wage rates, unemployment rates, labour surpluses) selected for inclusion in these mass variables can be projected via the use of input-output, econometric or combined input-output and econometric models -- and the nature of such integrated modelling efforts are discussed on pp. 180-185 and 238-241. The remaining variables (eg. climate) must be projected subjectively on the basis of past trends.

(3) Finally, where more than one factor is selected for inclusion in $^JO_{d\theta}$ (or $^LD_{d\theta}$) we need to establish demographic class specific weights for each factor in order to come up with the required composite index during the base year (and projections of such weights for the projection period if we believe they have changed since the base year). See pp. 174-180 for one means via which this problem could be resolved -- given sufficient data to run an econometric model of demographic class specific in- and out-migration flows.[43]

c. Doubly Constrained Models[44]

As mentioned above, one of the most difficult problems which needs to be resolved when adopting a gravity model approach is the identification of the weights to be employed in constructing the indices $^JO_\theta$ and $^LD_\theta$. (Even when one variable is selected to represent $^JO_\theta$ and another to represent $^L D_\theta$, this involves an implicit set of weights -- namely 1 (or 100 percent) for the variable selected, and O (or O percent) for all other variables originally considered for inclusion). One means by which this problem can (potentially) be resolved is to let $^J O_\theta$ = the number of net outmigrants originating from region J, and $^L D_\theta$ = the number of net inmigrants terminating in region L. We can then (data permitting) use a set of econometric equations of the form[45]

$$(^{J\rightarrow}M_\theta / ^J\underline{P}_\theta) = v_o + \sum_{i=1}^{n} v_i \, ^J X_{\theta i} + \mu_1$$

(5.3.17)

to both (1) establish the 'weights' for the pull-out factors $^J X_{\theta i}$ (i=1,...,n) which migration theory suggests are important for determining gross outmigration $^{J\rightarrow}M_\theta$, and (2) forecast gross outmigration flows $^{J\rightarrow}M_\Omega$, given projections of the $^J X_{\Omega i}$ derived from other modules (primarily input-output and econometric) comprising the integrated model, or from use of extrapolation based on past trends etc. as discussed above. In a parallel fashion, we can also (data permitting) use a set of econometric equations of the form[46]

$$(^{\rightarrow L}M_\theta / ^L\underline{P}_\theta) = U_o + \sum_{j=1}^{m} u_j \, ^L Y_{\theta j} + \mu 3$$

(5.3.18)

to both (1) establish the 'weights' for the push-out factors $^L Y_{\theta j}$ (j=1,...,m) which migration theory suggests are important for determining gross inmigration $^{\rightarrow L}M_\theta$, and (2) forecast gross inmigration flows $^{\rightarrow L}M_\Omega$, once again using projections of the $^L Y_{\Omega j}$ derived from other modules comprising an integrated model, or from extrapolation on the basis of past trends, and so on.

Econometric models of gross out- and in-migration flows will be discussed at length below, and so the exact form of equations (5.3.17) and (5.3.18) need not concern us here. Rather what needs to be recognized is that given forecasts of gross outmigration $\vec{^JM}_\Omega$ (J=A,...,U) and gross inmigration $\vec{^LM}_\Omega$ (L=A,...,U) from econometric equations of this form, we can easily derive forecasts of net outmigrations ($^JO_\Omega = \vec{^JM}_\Omega - {^J\vec{M}_\Omega}$) and net inmigrations ($^LD_\Omega = \vec{^LM}_\Omega - {^L\vec{M}_\Omega}$). Net interregional migration flows $^{JL}M_\Omega$ can then be forecast by substituting these $^JO_\Omega$ and $^LD_\Omega$ values into a gravity model of the form:[47]

$$^{JL}M_\theta = {^Ja}\, {^Lb}\, {^JO_\theta}\, {^LD_\theta}\, f(^{JL}d'') \qquad (5.3.19)$$

subject to

$$^JO_\theta = \sum_{L=A}^{U} {^{JL}M_\theta} = \vec{^JM}_\theta - \vec{^JM}_\theta \qquad (J \neq L) \qquad (5.3.20)$$

$$^LD_\theta = \sum_{J=A}^{U} {^{JL}M_\theta} = \vec{^LM}_\theta - \vec{^LM}_\theta \qquad (J \neq L) \qquad (5.3.21)$$

where ^Ja and ^Lb are balancing factors of the form

$$^Ja = \left\{ \sum_{L=A}^{U} {^Lb}\, {^LD_\theta}\, d(^{JL}s'') \right\}^{-1} \qquad (J \neq L)$$

and

$$^L b = \left\{ \sum_{J=A}^{U} {}^J a \, {}^J O_\theta \, f({}^{JL}d") \right\}^{-1} \qquad (J \neq L)$$

derived in an iterative manner during the calibration process.[48,49]

A similar set of econometric equations can be used (data permitting) to project gross out- and in-migration flows for each demographic class d. That is we may have:[50]

$$({}^{J\rightarrow}M_{d\theta}/{}^J P_{d\theta}) = v_{do} + \sum_{i=1}^{n} v_{di} \, {}^J X_{di\theta} + \mu_5 \qquad (5.3.22)$$

and

$$({}^{\rightarrow L}M_{d\theta}/{}^L P_{d\theta}) = u_{do} + \sum_{j=1}^{m} u_{di} \, {}^L Y_{di\theta} + \mu_6 \qquad (5.3.23)$$

Then letting ${}^J O_{d\theta}$ = the number of net outmigrants of demographic class d originating from region J, and ${}^L D_{d\theta}$ = the number of net inmigrants terminating in region L we then employ a modified gravity model of the form:[51]

$$^{JL}M_{d\theta} = {}^J a_d \, {}^L b_d \, {}^J O_{d\theta} \, {}^L D_{d\theta} \, f({}^{JL}d") \qquad (5.3.24)$$

subject to

$$^J O_{d\theta} = \sum_{L=A}^{U} {}^{JL}M_{d\theta} = {}^{L\rightarrow}M_{d\theta} - {}^{\rightarrow J}M_{d\theta} \qquad (L \neq J)$$

$$(5.3.25)$$

$$^LD_{d\theta} = \sum_{J=A}^{U} {}^{JL}M_{d\theta} = {}^{L\rightarrow}M_{d\theta} - {}^{\rightarrow L}M_{dq} \qquad (L \neq J)$$

(5.3.26)

where ^Ja_d and ^Lb_d are once again conversion constants derived iteratively during the calibration process, and where $^{J\rightarrow}M_{d\Omega}$ and $^{\rightarrow J}M_{d\Omega}$ (J=A, ..., U) are derived from the econometric equations (5.3.22) and (5.3.23) during the projection period. See pp. 174-180 for a discussion of econometric equations for demographic class specific gross out- and in-migration flows.

d. Other Constrained Gravity - Type Models[52]

One problem with the gravity model formulation just outlined is that it requires historical data on origin-specific, destination-specific, net migration flows $^{JL}M_\theta$ (J=A,...,U; J=L) or $^{JL}M_{d\theta}$ (d=1,...,d; J=A,...,U; J≠L). However, where such data is not available (or where none of the econometric equations to be discussed below produce reasonable results) we may resort to the use of a non-behavioral technical-type of gravity model formula. This 'last-resort' approach would include the following steps:

Step 1. First, we employ an interregional input-output submodel (or a CICIOP type module) to derive a projection of output by sector by region. We then use a set of fixed output to employment coefficients to derive corresponding projections of employment demand by demographic class $_D{}^JE_{d(t+\theta)}$ (d=1,...,d; J=A,...,U).

Step 2. Next, we simply age the existing population (or use someone else's projections of such natural increase in each region's population)[53] and add in a reasonable level of net overseas migration to yield a crude set of regional population estimates excluding the effects of interregional migration. We then use a set of fixed labour force participation rates to yield projections of labour supply by demographic class $^{SJ}E_{d(t+\theta)}$ (d=1,...,d; J=A, ..., U).

Step 3. Using the employment demand and labour supply estimates from Steps 1 and 2, we next derive estimates of the surplus $^{J}S_{d(t+\theta)}$ or deficit $^{L}D_{d(t+\theta)}$ labour in each region. Net interregional migration flows $^{JL}M_{d\theta}$ (d=1,...,d; J=A,...,U; J=L) are then determined as

$$^{JL}M_{d\theta} = g\, ^{J}S_{d(t+\theta)}\, ^{L}D_{d(t+\theta)}\, (^{JL}d")^{-1} \qquad (5.3.27)$$

where g is an arbitrary conversion factor (with distance times (time/persons) as its unit).

Step 4. Next, a number of constraints need to be imposed on the results of equation (5.3.27). Firstly, we require that all regional deficits of labour are eliminated. That is:

$$\vec{\,}^{L}M_{d\theta} \geq {}^{L}D_{d(t+\theta)}$$

where (5.3.28)

$$\vec{\,}^{L}M_{d\theta} = \sum_{J=A}^{U} {}^{JL}M_{d\theta} \qquad (J \neq L)$$

If such is not the case for region L, we must introduce deficit region specific factors $^{L}\alpha$ to adjust the g constant upwards to the point where $\vec{\,}^{L}M_{d\theta} = {}^{L}D_{d(t+\theta)}$.[54]

Secondly, we must also check that the net outmigration from any given region does not exceed its projected labour surplus. That is, it must be that:

$$^{J\rightarrow}M_{d\theta} \leq {}^{J}S_{d(t+\theta)}$$

where (5.3.29)

$$^{J\rightarrow}M_{d\theta} = \sum_{L=A}^{U} {}^{JL}M_{d\theta} \quad (J \neq L)$$

If such is not the case for region J, we must introduce surplus region specific factors $^J\beta$ to adjust the g constant downwards to the point where $^{J\rightarrow}M_{d\theta} = {}^{J}S_{d(t+\theta)}$. The formula then becomes:

$$^{JL}M_{d\theta} = G \, {}^{L}\alpha \, {}^{J}\beta \, {}^{J}S_{d(t+\theta)} \, {}^{L}D_{d(t+\theta)} \, ({}^{JL}d")^{-1} \quad (5.3.30)$$

In general, we may need several rounds of adjustments to achieve convergence to a solution in which all deficits are met without exceeding the surplus of any region. We may also impose other constraints to ensure that the formula in (5.3.30) yields reasonable results. For example, one set of constraints might be of the form $(^{K}UN_{t+\theta}/{}^{K}\underline{P}_{t+\theta}) \geq$, $= k$ (K=A, .., U) where $^{K}UN_{t+\theta} = {}^{K}S_{d(t+\theta)} - {}^{K\rightarrow}M_{d\theta}$ for surplus regions and $^{\rightarrow K}M_{d\theta} - {}^{K}D_{d(t+\theta)}$ for deficit regions and k, is the maximum acceptable unemployment rate for any region. Alternatively, we may have $(^{J}UN_{t+\theta}/{}^{J}\underline{P}_{t+\theta}) - (^{L}UN_{t+\theta}/{}^{L}\underline{P}_{t+\theta}) \leq q$ (J,L=A,...,U; J≠L) where q is the maximum acceptable discrepancy in unemployment rates between any pair of regions, based upon historical data and other information pertaining to future policies.

5.3.2 Econometric Approach

Although there is no one econometric model which has widespread acceptance for use in forecasting interregional migration flows, this approach has nevertheless been widely used. (For example, see Harris (1973, 1980), Olsen, et al (1977), Milne, Glickman and Adams (1980), Milne (1981), Cebula (1979), Todaro (1976) and Fields (1976)). A number of reasons can be cited as contributing to this lack of agreement.

(1) Firstly, analysts have employed different measures of migration. For example, some attempt to model net migration (Harris (1973, 1980), Milne, Glickman and Adams (1980), Milne (1981) and Cebula (1979)); while others model gross migration flows (Olsen, et al (1977) and Greenwood (1981)).[55]

(2) Secondly, analysts have adopted different levels of aggregation in their measures of migration. For example, some attempt to construct a different model for each of a number of demographic classes (Gober-Meyers (1978), Cebula (1979, Chapters 2 and 5), Milne, Glickman and Adams (1980) and Milne (1981)); while others employ just one model, namely one for the population as a whole (Harris (1973, 1980), Olsen, et al (1977) and Cebula (1979, Chapter 3)).[56]

(3) Thirdly, analysts have employed different levels of regional aggregation -- eg. some have defined regions in such a way as to emphasise rural-to-urban migration flows (Todaro, 1976),[57] others in such a way as to emphasise interstate migration flows (Gober-Meyers (1978) and Wadycki (1979)), and yet others in such a way that commuting needs to be considered as an alternative mechanism via which a region can achieve the required balance between the supply of and demand for labour (Birg (1981) and van der Veen and Evers (1982)).

(4) Fourthly, a state of agreement has not yet been reached concerning the major economic determinants of interregional migration,[58] nor on the relative importance of economic and non-economic determinants.[59] As a result, the explanatory variables

included in econometric models reflect (at least in part) the analyst's preferences for the competing theories available.[60]

(5) Fifthly, any given 'theory of migration' must be translated into a form capable of implementation given available data. Not only does the data available to analysts differ, but so do analysts' choices as to the most appropriate proxy for a given determinant should the most directly relevant variable be unavailable.[61] Additionally, the theory is often vague in terms of (a) the precise nature of a particular determinant[62], and/or (b) the precise nature of the time lag involved between the experience of a given stimulus (determinant) and the migration response.[63] These last two issues must be resolved by the analyst, and any two analysts may be expected to differ somewhat in the manner in which they choose to resolve them.

(6) Sixthly, some analysts believe that migration flows can be adequately modelled by employing a single equation approach (Cebula (1979), Levy and Wadycki (1974), Wadycki (1979) and Feder (1979, 1980)). Others however believe that a system of equations is required (Alperovich, Bergsman and Ehemann (1977), Olsen, et al (1977), Gober-Meyers (1978) Kau and Sirmans (1979) and Greenwood (1981)). Here again the analyst's judgement plays a role in determining the system of equations he considers appropriate. For example, some might first set up an equation for determining migration flows out of each region (thus emphasising the 'push-out' factors in determining a pool of migrants; secondly, set up a share equation for allocating the outmigrants from any region among the remaining regions in the system in proportion to their relative attractiveness (thus emphasising the role of 'pull-in' factors in the choice of a destination region); and finally, set up an equation for determining the total migration flows into each region (as an identity employing the results of the above two sets of equations). On the other hand, some might reverse the order of the operations described above -- first determining regional inflows of migrants, secondly

determining the regional origins of these inflows on the basis of a share equation, and finally deriving regional outflows of migrants as an identity.

(7) Finally, analysts differ in the extent to which they attempt to 'correct' the model when either (a) the assumptions underlying the ordinary least squares technique are violated; or (b) the results derived from implementing the model are not considered reasonable.[64]

As a result of these factors, it does not seem fruitful to include an extensive review of the literature. Rather, given (a) the experience (successes and failures) of others, and (b) discussions with colleagues on what are the forces driving interstate migration flows in the Australian context, I have chosen to develop the following as a set of econometric equations which both seem reasonable to explore and can be implemented with available data.[65]

a. Inmigration Equations

For those demographic classes d which include members of the labour force aged 15-64 we have:

$$({}^L M_d / {}_S{}^L E_d) = a_0 + a_1 [({}^L W Y_d / {}_D{}^L E_d)/(W Y_d / {}_D E_d)]$$
$$+ a_2 [(\Delta_D {}^L E_d / {}_D{}^L E_d)/(\Delta_D E_d / {}_D E_d)]$$
$$+ a_3 [({}_D{}^L E_d / {}_S{}^L E_d)/({}_D E_d / {}_S E_d)] + \mu_1 \qquad (5.3.31)$$

where ${}^L M_d$ = migration into region L of labour force in demographic class d;

${}_S{}^L E_d$ = labour force (supply) in demographic class d in region L;

${}_D{}^L E_d$ = employment demand for members of demographic class d in region L;

${}^L W Y_d$ = wage income accruing to members of demographic class d in region L; and

μ_1 = a stochastic error term.

The second term in (5.3.31), namely the wage income per employee in region L relative to the nation, is included as a proxy for the potential gains from migration out of L.[66] The third term, the change in employment demand in region L compared with the nation, is included as a proxy for the change in employment opportunities in L relative to elsewhere. The fourth term, the ratio of employment demand relative to labour supply in region L compared with the nation, is included as a proxy for the probability of obtaining employment in L relative to elsewhere.

There are several problems involved with modelling labour force inmigration in this way:

(1) The first issue involves the question of timing. The migration flows take place between two points of time, namely t-5 and t, so that one has a 'choice' as to which year the 'at risk' population in the dependent variable of equation (5.3.31) refers -- is it t-5 or t, or some intermediate point? In order to keep things simple we propose to take the unweighted average of the 'at risk' population in years t-5 and t. Similarly, while the change components of the independent variables of equation (5.3.31) may be taken to refer to changes in employment demand between t-5 and t, the other components can once again refer to conditions at either t-5 or t or some point intermediate between t-5 and t. This timing problem is discussed in more detail in Smith (1983,pp.300-301) and so will not elaborate further here except to propose the use of a simple average of conditions between t and t-5 (or conditions at year t-3) wherever such a problem arises.

(2) The second issue was discussed at length on pp. 166-167, and concerns the question of whether people consider just the 'benefits' from migration accruing to members of their present demographic class or whether they take into account the possibility of changes in their demographic class affiliation over their lifetime. Once again, we introduce the latter type of calculation by modifying (5.3.31) to yield

$$(^{\rightarrow L}M_d/_S\,{}^LE_d) = c_0 + c_1 \sum_{d'} q_{dd'} \left[(^L WY_{d'}/_D\,{}^LE_{d'}/_DE_{d'})\right]$$

$$+ c_2 \sum_{d'} q_{dd'} \left[(\Delta_D\,{}^LE_{d'}/_D\,{}^LE_{d'})/(\Delta_D E_{d'}/_D E_{d'})\right]$$

$$+ c_3 \sum_{d'} \left[(_D\,{}^LE_{d'}/_S\,{}^LE_{d'}/_S\,{}^LE_{d'})/(_D E_{d'}/_S E_{d'})\right] + \mu_3$$

(5.3.32)

where d'ε D and D is the set of demographic classes which persons in demographic class d could reasonably expect to reach during their lifetime, and yet are not so 'distant' as to be eliminated from 'present value' type of evaluations of migration gains; while $q_{dd'}$ is the weight attached to conditions prevailing for members of d' by current members of d.

(3) The third issue relates to data problems. Since internal migration questions have only been asked on the last four population censuses (1971, 1976, 1981 and 1986) in Australia, we must pool time series and cross section data in order to be able to derive meaningful results from an econometric model. Such does not permit the identification of regional differences in inmigration relationships, although one or more regional dummies may be introduced to get around this problem somewhat.[67] This pooling of data also provides a rationale for use of the dependent variable ($^{\rightarrow L}M_d/_S{}^LE$) rather than $^{\rightarrow L}M_d$ in equations (5.3.31) and (5.3.32). This is done to avoid the heteroscedasticity problem which would otherwise arise because of regional differences in population size, and hence in inmigration absorbing potential.[68]

b. Outmigration Equations

A similar set of econometric equations is proposed for modelling gross regional outmigration $^{J\!\!\rightarrow}M_d$ by members of the labour force aged 15-64. That is:

$$(^{J\!\!\rightarrow}M_d/_S{}^J E_d) = e_0 + e_1 \sum_{d'} q_{dd'} \left[(^J WY_{d'}/_D{}^J E_{d'})/WY_{d'}/_D E_{d'}) \right]$$

$$+ e_2 \sum_{d'} q_{dd'} \frac{[\Delta(^J WY_{d'}/_D{}^J E_{d'})/(^J WY_{d'}/_D{}^J E_{d'})]}{[\Delta(WY_{d'}/_D E_{d'})/(WY_{d'}/_D E_{d'})]}$$

$$+ e_3 \sum_{d'} q_{dd'} \frac{[(_S{}^J E_{d'} - _D{}^J E_{d'}/_S{}^J E_{d'}]}{[(_S E_{d'} - _D E_{d'})]} + \mu_4 \quad (5.3.33)$$

Thus gross outmigration 'per capita' is hypothesised to increase (a) as wage rates decline relative to the nation (the second term on the RHS of (5.3.33)), (b) as the percent increase in wage rates decline relative to the nation (the third term), and (c) as unemployment rates increase relative to the nation (the fourth term).[69]

A parallel set of problems arise with respect to this formulation, that is, the question of timing arises, as does the question of the set of demographic classes D to which the RHS variables in (5.3.33) refer. Once again regional differences have to be ignored since we must pool time series and cross sectional data, and the dependent variable is expressed in 'per capita' terms to get around the heteroscedasticity problem which would otherwise arise because of regional differences in population size and hence outmigration generating potentials.[70]

c. Interregional Migration Flows

We propose to model interregional migration flows for members of the labour force aged 15-64 via the use of a doubly constrained gravity model of the form given by equations (5.3.24)-(5.3.26) above. During the projection period the required outmigration $^J\!\!\Rightarrow\! M_{d\Omega}$ and inmigration $\Rightarrow\! ^LM_{d\Omega}$ estimates will be derived from the set of econometric equations just described.

Interregional net migration rates for (a) demographic classes $\underline{d}$ comprising persons aged 15-64 who are not members of the labour force, and (b) demographic classes d! comprising persons aged 0-14, are assumed to change over time in the same proportion as those for members of the labour force.[71] This then leaves the interregional migration patterns of persons aged 65 and over to be determined. We propose to do this via the use of a set of fixed destination-specific migration rates for each region.[72,73]

5.4 Integration of Multiregional Input-Output and Demographic Models

From a theoretical standpoint, forecasts of regional population change cannot be considered sound if they ignore the effects of socio-economic conditions on interregional migration flows. Likewise, models of regional economic change must be considered incomplete if they ignore the effects of population change on socio-economic conditions. As a result the linkage of demographic and economic models has been suggested by a number of analysts -- for example, Courbis (1975), Ballard and Wendling (1980), Harris (1973, 1980), Birg (1981), Beaumont, et al (1982), Isserman (1980, 1982), Ledent (1978, 1982), Olsen, et al (1977), Milne, et al (1980), and Milne (1981). Recently a number of approaches have been developed for achieving this linkage using a regional or interregional input-output model -- for example, Schinnar (1976, 1977), Evans and Baxter (1980), Dulgeroff (1980), Hollenbeck (1980), Beyers (1980), Gordon and Ledent (1980, 1981), Madden and Batey (1980), Batey and Madden (1980, 1983), Betson (1981) and Isard and Smith (1982n). A schematic representation of these integrated approaches is given in Figure 5.1. The basic steps involved are:

Step 1: *Derivation of Crude Population Projections*

The first step involves the aging of the base year population $^J\underline{P}_t$ or $^J\underline{P}_{dt}$ (d=1,...,d) in each region J (J=A,...,U) to give an estimate of population $^J\underline{P}_{t+\theta}$ or $^J\underline{P}_{dt+\theta}$ (d=1,...,d), excluding the effects of foreign and interregional migration over the projection period θ. This involves estimating the number of births and deaths expected for the population, as well as non-interregional movements from one demographic class to another over the projection period. This can be done using one of the procedures outlined on pp. 149-160.

Step 2: *Adjustment for Non-Economically Induced Migration*

The second step involves the adjustment of the population estimates $^J\underline{P}_{t+\theta}$ or $^J\underline{P}_{d(t+\theta)}$ (d=1,...,d) derived in Step 1 for the level of migration which is assumed not to be dependent on prevailing regional and/or national economic conditions. That is, we add in the levels of:

(1) net foreign immigration, using the procedure outlined on pp. 151-152 above for allocating a given national total among the U regions; and

(2) interregional migration of persons, aged 65 and over, using the fixed set of destination - specific rates for each region discussed on page 180.

Step 3: *Derivation of Initial Population-Sensitive Final Demand Projections*

The third step involves the use of the adjusted population estimates (excluding interregional migration) derived at the end of Step 2 to get an initial final demand estimate for those components of final demand

182 CHAPTER 5

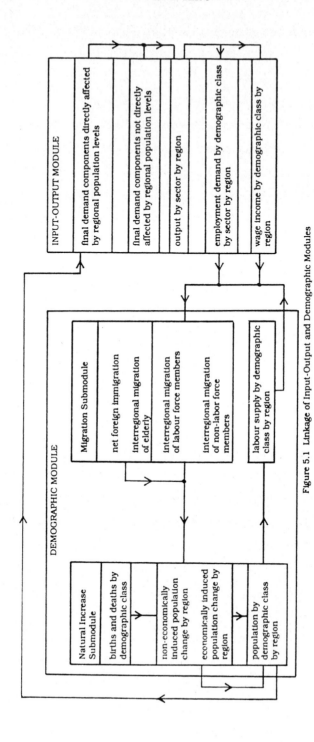

Figure 5.1 Linkage of Input-Output and Demographic Modules

which are directly dependent on regional population levels.[74] (These include, for example, household consumption and some government programs.)

Step 4: *Derivation of Remaining Final Demand Projections*

The fourth step involves the crude estimation of the level of those components of final demand which are not directly dependent on regional population levels.

Step 5: *Derivation of Initial Output Projections*

The fifth step involves the use of the final demand estimates derived in Steps 3 and 4 to get an initial estimate of output levels by sector by region. This is done via the use of an interregional input-output module of the type described on pp. 17-35.

Step 6: *Derivation of Initial Labour Requirements Projections*

The next step involves the use of the output estimates derived in Step 5 to get an initial estimate of employment demand (labour requirements) in each region for year $t+\theta$. This is done by the use of a set of 'reasonable' output to employment coefficients for each sector in each region. These coefficients can relate to total labour requirements or be broken down by demographic class depending on the level of disaggregation required for the operation of the migration component of the demographic model.

Step 7: *Derivation of Crude Wage and Salary Income Projections*

The seventh step involves the derivation of crude estimates of regional wage and salary incomes for the year $t+\theta$. Where the migration component of the demographic model does not require demographic class specific wage and salary income projections, then this is done crudely via the use of standard input-output procedures. That is, by multiplying each sector's output level $^J X_i$ derived in Step 5 by the dollars worth of labour required per dollar output (i.e. by the $^J a_{Li}$ coefficient) derived from the base-year input-output table. The resulting wage and salary incomes are then summed over all sectors. That is: $^J WY = \sum_i {^J a_{Li}} {^J X_i}$.

However, when regional wage and salary income projections are required for each demographic class then a set of demographic class specific wage rates must be projected for each region -- perhaps by assuming they have remained unchanged from their base-year counterparts. These wage rates are then multiplied by the employment demand projections derived in Step 6, to give the required set of wage and salary income projections. These projections are then summed over all demographic classes to yield a total wage and salary income projection $^J WY'$ for each region. Should major inconsistencies arise between the $^J WY$ and $^J WY'$ projections, they can be eliminated through the use of a series of proportional adjustments.

Step 8: *Derivation of Initial Labour Supply Projections*

The next step involves the use of the population estimates derived at the end of Step 2 to get initial crude labour supply projections (excluding the effects of economically-induced interregional migration flows) for year $t+\theta$. This is done via the use of a set of relevant (crudely estimated) labour force participation rates for each region.[75]

Step 9: *Derivation of Gross Outmigration and Gross Inmigration Rate Projections for Each Region's Labour Force*

The results of Step 6 (regional employment demands), Step 7 (regional wage and salary incomes) and Step 8 (regional labour supply) can then be used as RHS variables in the migration equations described on pp. 176-179. This yields projections of gross inmigration and gross outmigration rates for each region's labour force over the projection period.

Step 10: *Estimation of Labour Force 'At Risk' of Migration Over Projection Period*

The next step involves the estimation for each region of the labour force 'at risk' of migration over the projection period. This is done crudely by simply averaging the base-year labour supply $_S{}^J E_t$ (or $_S{}^J E_{dt}$) with that estimated in Step 8 as the labour force level $_S{}^J E_{t+\theta}$ (or $_S{}^J E_{d(t+\theta)}$) at the end of the projection period.

Step 11: *Projection of Gross Labour Force Outmigration and Inmigration Levels for Each Region*

Multiplication of the migration rates (derived in Step 9) by the corresponding 'at risk' labour force (derived in Step 10) yields estimates of the level of gross outmigration and gross inmigration for each region's labour force.

Step 12: *Projection of Net Interregional Labour Force Migration Flows*

The results of Step 11 can now be used in the gravity model described on pp. 169-173 to yield an estimate of net interregional migration flows by members of the labour force aged 15-64 over the projection period.

Step 13: *Derivation of Net Interregional Migration Flow Projections for Non-Labour Force Members Aged 15-64*

The next step involves the projection of net interregional migration rates for persons aged 15-64 who are not members of the labour force. This involves:

(a) the use of the results of Steps 10 and 12 to derive an estimate of net interregional migration rates for members of the labour force over the projection period;

(b) the use of these labour force migration rates in equation (5.3.34) to adjust the base-year interregional migration rates of non-labour force members aged 15-64;

(c) the use of the results of Steps 2 and 8 above and a simple averaging procedure similar to that described in Step 10 to derive, for each region, an estimate of non-labour force members aged 15-64 who were 'at risk' of migrating between years t and t+θ; and

(d) multiplication of the migration rates derived in (b) by the 'at risk' population levels derived in (c).

Step 14: *Derivation of Net Interregional Migration Flow Projections for Non-Labour Force Members Aged 0-14*

This is done using a procedure almost identical to that just described in Step 13, using equation (5.3.35) above to adjust the base-year interregional migration rates for persons aged 0-14.

Step 15: *Consolidation of Net Interregional Migration Projections*

The results of Steps 12, 13, and 14 can now be combined to yield an estimate of the economically induced changes in each region's population over the projection period. These estimates are then used to revise the initial crude population projections derived at the end of Step 2.

Steps 3-15 are then repeated as many times as is required in order to converge on a reasonably consistent set of results.

5.5 *Integration of Multiregional Input-Output, Comparative Cost-Industrial Complex, Programming and Demographic Modules*

A schematic representation of the nature of this more complete integration is given in Figure 5.2. The basic steps involved are:

DEMOGRAPHIC MODULE

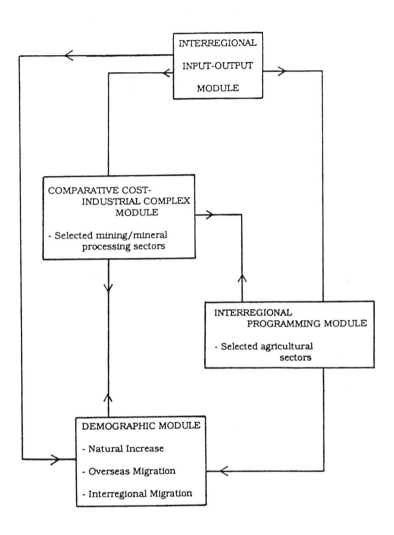

Figure 5.2 Linkage of Input-Output, Comparative Cost-Industrial Complex, Programming and Demographic Modules

Step 1: *Derivation of Crude Population Projections*
As in Step 1 of Section 5.4.

Step 2: *Adjustment for Non-Economically Induced Migration*
As in Step 2 of Section 5.4.

Step 3: *Identification of Sectoral Structure and Setting Up of Modified Interregional Input-Output Matrix*

Each region's economy is divided into four sets of sectors: energy supply sectors S, energy product sectors P, cost-sensitive non-energy sectors N', and other non-energy sectors N". The cost-sensitive sectors are removed from the interindustry section of the interregional input-output table and placed within a set of 'non-traditional' final demand columns and primary input rows and the modified interregional technical-trade coefficients matrix partitioned in the manner required for joint operation of the interregional input-output and programming models.

Step 4: *Derivation of Initial Traditional Final Demand Projections*

The next step involves the projection of an initial set of traditional final demand vectors $Y' = [Y'_{N'}\ Y'_{N''}\ Y'_P\ Y'_S]$ for year $t+\theta$. Exogenous projections are first identified for the following system (national) variables: output by sector, employment by sector, and final demand by type by sector. These projections should be consistent with the analyst's basic assumptions about the future of the system -- including the natural increase and immigration levels estimated in Steps 1 and 2 above. (They would typically come from the operation of existing national (system) input-output and econometric models). These national final demand estimates are then disaggregated by region, and adjusted for net foreign imports to yield a projection of the level of 'traditional' final demand by sector by region.

Step 5: Derivation of Market Projections for Cost-Sensitive Non-Energy Sectors

The next step involves the estimation of the level of regional demands (intermediate and final) for the output of each cost-sensitive non-energy sector in the year $t+\theta$. (The final demand projections Y_N' were estimated in Step 4, so essentially this step involves projection of the requirements for these sectors' outputs as intermediates).

Step 6: Derivation of Initial Non-Traditional Final Demand Projections

Given the pattern of markets identified in Step 5, the next step involves the use of the multiregional variants of the comparative cost and/or industrial complex approaches to derive projections for each region of the level of output produced by each industry (or part of industry) placed within the non-traditional final demand column(s) of the input-output module. See pages 54-60 for further details.

Step 7: Consolidation of Traditional and Non-Traditional Final Demand Projections

Now that the levels of 'traditional' and 'non-traditional' final demand by sector by region have been projected for the year $t+\theta$ (in Steps 4 and 6, respectively), they can be consolidated into the combined traditional and non-traditional final demand vectors Y_N', Y_N'', Y_P, and Y_S required for operation of the input-output and programming modules.

Step 8: Derivation of Initial Output Projections for Non-Energy and Energy Product Sectors

Using the results of Steps 3 and 7, as in Step 8 of Section 4.4.2.

Step 9: *Derivation of Functional Energy Service Demand Level Projections*

The output estimates X_P derived in Step 8 are used to provide a first approximation of regional demands for energy end-products. By employing a set of aggregation and/or disaggregation procedures, these estimates can be converted into the functional energy service demand levels $^J D_d$ (d=1, ..., 17) required as binding magnitudes for the demand constraints within the linear programming module.

Step 10: *Derivation of Initial Output Projections for Energy Supply Sectors*

Using the results of Steps 7-9, as in Step 10 of Section 4.4.2.

Step 11: *Successive Reruns of the Input-Output and Programming Modules*

The results of Step 10 are now used to derive revised estimates of X_N'', X_P and $^J D_d$ (J=A,...,U; d=1,...,17). The linear programming module is rerun, and new projections of X_S derived. And so on, with Steps 8-11 repeated until a point is reached where the input-output and programming modules converge on a stable set of outputs for the energy product P, energy supply S and other non-energy N" sectors.

Step 12: *Identification of Inconsistencies in Output Projections*

The set of output vectors X_N', X_N'', X_P and X_S derived from a first run through of steps 3-11 can generally be expected to give rise to a number of inconsistencies of the type described in Step 12 of Section 4.4.2.

Step 13: *Elimination of Major Inconsistencies*

If any of the inconsistencies identified in Step 12 are judged to be too large to be permitted in the final projections, re-estimations of the type described in Step 13 of Section 4.4.2 will need to be performed. The results of Step 11 provide the basic information to be employed in this re-estimation process.

Steps 6-13 are then repeated, and a new set of output estimates derived for each region/sector. These results are checked for inconsistencies. And so on, until all major discrepancies have been eliminated.

Step 14: *Derivation of Initial Employment Demand Projections*

As in Step 6 of Section 5.4, using the output projections derived at the end of Step 13 above.

Step 15: *Derivation of Crude Wage and Salary Income Projections*

As in Step 7 of Section 5.4, using the results of Steps 13 (output by sector) and 14 (regional employment demands).

Step 16: *Derivation of Initial Labour Supply Projections*

As in Step 8 of Section 5.4, using the results of Step 2 (crude poulation estimates).

Step 17: *Derivation of Gross Outmigration and Gross Inmigration Rate Projections for Each Region's Labour Force*

As in Step 9 of Section 5.4, using the results of Steps 14 (regional employment demands), 15 (regional wage and salary incomes) and 16 (regional labour supplies).

Step 18: *Estimation of Labour Force 'At Risk' of Migration over the Projection Period*

As in Step 10 of Section 5.4, using the results of Step 16 above.

Step 19: *Projection of Gross Outmigration and Gross Inmigration Levels for Each Region*

As in Step 11 of Section 5.4, using the results of Steps 17 and 18 above.

Step 20: *Projection of Net Interregional Labour Force Migration*

As in Step 12 of Section 5.4, using the results of Step 19 above.

Step 21: *Projection of Net Interregional Migration Flows for Non-Labour Force Members*

As in Steps 13 and 14 of Section 5.4, using the results of Steps 2, 16, 18 and 20 above.

Step 22: *Consolidation of Net Interregional Migration Projections*

The results of Steps 20 and 21 can now be combined to yield an estimate of the economically induced change in each region's population over the projection period. These estimates are then used to revise the initial crude population projections derived at the end of Step 2.

Steps 3-22 are then repeated as many times as is required in order to converge on a reasonably consistent set of results.

5.6 Concluding Remarks

This chapter has dealt with the demographic module and its linkage to the input-output, comparative cost-industrial complex and linear programming modules of an integrated multiregion model. The module proposed for use in the Australian context includes:

(a) a set of econometric equations for determining regional out- and in-migration flows; and

(b) a doubly-constrained gravity model for determining the pattern of interregional migration flows -- given the control totals identified in (a).

It has been specified in such a way that the data are available to support its implementation. All that is required is that the necessary census data on interregional migration flows be requested from the Australian Bureau of Statistics, and an appropriate computer program be written.

However, should the econometric equations (a) give unreasonable parameter values (be insignificant, have wrong signs, etc.), or (b) have inadequate explanatory power for use in projecting future interregional migration flows, then the analyst can always make use of the fall-back gravity model proposed as a last-resort on pp. 171-173.

NOTES

[1] However, the conceptual limitations and poor track record of this approach is well documented in the literature. See, for example, Siegel (1972), Lee (1976), and Isserman (1982).

[2] This is especially difficult where no clear-cut pattern of regional convergence (divergence) can be identified. For example, see O'Connell (1981). However, even when regional fertility patterns appear to be converging around the national average over time, the problem remains of forecasting the rate of convergence. For example, see Olsen et al (1977).

[3] See Isserman (1980, 1982) for an overview of the growing literature on socio-economic determinants of fertility. A number of studies (e.g. Gregory, et al (1972), Wachter (1975) and Butz and Ward (1979)), have attempted to incorporate some of the more widely accepted of these determinants into models of national fertility. In addition, some more recent studies (e.g. Ahlburg (1982) and Redwood (1982)) have taken the additional step of forecasting state fertility on the basis of these more theoretically sound national projections.

[4] Note that where $J=L$, we have the net proportion of region L's population at time t who stay within the region through time $t+\theta$.

[5] Note that the relationships in (5.2.2) and (5.2.3) represent only crude estimates of regional population change due to natural increase and migration respectively. This is so because it needs to be recognized that (a) some of the infants born during period θ will either (1) die, or (2) migrate out of region J before year $t+\theta$; and (b) some of the inmigrants to region J during period θ will either (1) give birth, (2) die, or (3) both give birth and die before year $t+\theta$. The interested reader may refer to Rees and Wilson (1973, 1975 and 1977), Rees and Willekens (1981) and Beaumont, Isserman, et al (1982) for a detailed description of how these adjustments can be made.

[6] For an example of such an approach, see Olsen et al (1977).

[7] It may be argued that the interregional allocation of migrants should be allowed to vary as a function of economic conditions (say growth of employment opportunities) in the different regions. In fact, this may be a fruitful area for future research, especially in countries such as Australia which rely so heavily on overseas migration as a source of population growth. However, a preliminary examination of past trends (Rowland, 1979) suggests that migrants of a particular nationality tend to settle, initially at least, in regions with a large concentration of residents of the same national origin. This appears to be a way of alleviating the language, cultural and other problems associated with

settling in a new country -- and is a phenomena that is likely to continue. Hence, the use of past trends may yield reasonable projections -- especially when it is recognized that once these migrants enter the system their internal migration patterns are then taken to be identical to those of the existing population. Thus we are allowing foreign immigrants to respond to internal economic conditions, but hypothesising that such responses develop after a lag of one period.

[8] The matrix formulation of this technique owes much to the work of Rogers (1966, 1973, 1975) to whom the reader is referred for a more complete presentation.

[9] For example, if $\theta = 5$ and the last group corresponds to persons aged 65+, then $d = 14$.

[10] That is, $b_{d\theta}$ is defined as the ratio of the births from females in the age group d (during a period of θ years) over the total size of that group (inclusive of males).

[11] Even in this row, however, the entries will be non-zero only for those age groups corresponding to childbearing years for females.

[12] That is $s_{d\theta} = (1-d_{d\theta})$, where $d_{d\theta}$ is the death (mortality) rate of age group d in region J.

[13] Note the convention regarding the age group subscripts in this matrix -- in general, the survivorship ratio $s_{d\theta}$ appears in column d and row d+1.

[14] That is, $^{LJ}m_{d\theta} = {}^{LJ}M_{d\theta}/{}^{L}P_{dt}$ where $^{LJ}M_{d\theta}$ is the level of net migration from region L to region J by persons in age group d during period θ. Hence, $^{LJ}m_{d\theta} < 0$ when there is net inmigration from L, and $^{LJ}m_{d\theta} < 0$ when there is a net outmigration.

[15] Note that the convention regarding the age group subscripts in this matrix is the same as that used in the survivorship operator -- in general, the net migration rate $^{LJ}m_{d\theta}$ appears in column d and row d+1.

[16] That is, its first row comprises birth rates $^{J}b_{d\theta}$ (d=1,...,d), its lower left hand element and the elements directly below the main diagonal now comprise survivorship ratios $^{J}s_{d\theta}$ (d=1, ..., d) and all other elements are equal to zero.

[17] See Rodgers (1975), Rodgers and Williams (1982) and Rees and Wilson (1977) for details of this augmentation process.

[18] In addition, there is the need to correct for the lack of interdependence between demographic processes in the model. That is, we need to take into account (a) individuals born in region J during period θ who either die or migrate out before year t+θ; (b) individuals who migrate into region J during period θ and either give birth, or die, or both give birth and die before year t+θ, and so on. Demographers have developed means via which the necessary adjustments can be made. See Rees and Wilson (1975, 1977), Wilson and Rees (1976), Beaumont, Isserman, et al (1982), and Rees and Willekens (1981) for details.

[19] See Olsen, et al (1977), Ahlburg (1982) and Redwood (1982) for examples of the projection of these regional differences for a number of age-sex groups in the United States.

[20] Indeed this lack of responsiveness is the major drawback of these methods. For a discussion of other problems, see Shryock, Siegel et al (1976) and Plane and Rogerson (1982).

[21] The ideas underlying microsimulation (as the term is used here) were first introduced by Guy Orcutt nearly 25 years ago. (Orcutt, 1957). In developing this section, we draw heavily upon Orcutt et al (1976), Caldwell (1980, 1982) and Caldwell and Saltzman (1981). For some recent applications of the use of microsimulation models for the analysis of public policy, see Golliday and Haveman (1977) and Haveman and Hollenbeck (1980).

[22] This is so since by comparison with the enormous number of transition probabilities required for a transition matrix model, the decision unit probabilities required by a comparably detailed microsimulation model are relatively few. See Orcutt et al (1961) for further discussion.

[23] This is the approach adopted in the DYNASIM model. See Orcutt, et al (1976, Chapter 7).

[24] For example, see Spilerman (1972) and Juster and Land (1981) for attempts to treat transition probabilities as variables functionally dependent on attributes and/or environmental conditions supplied exogenously.

[25] An interesting area of research, however, would involve an attempt at integrating micro- and macro- approaches to migration modelling. For example, the base year micro data could be used to derive initial region-to-region migration probabilities. A macro-type of migration model could then be used to project such migration probabilities for the population as a whole, or for a limited set of population classes (eg. for

each age-sex group). The results of this macro-model could then be used to suggest required adjustments in the initial micro data-based probabilities.

[26] Another approach not covered in this book is the ratio method. The general procedure in this approach is to relate changes in net migration rates (or levels) to changes in some indicator of relative regional economic, social and/or political conditions considered attractive by persons making migration decisions. For example, when aggregate population projections are being made for each region we have a set of equations (for $J,L = A,...,U; J \neq L$) of the form:

$$^{LJ}m_\Omega = {}^{LJ}m_\theta \, (^J\!A_\Omega / \sum_{J=A}^{U} {}^J\!A_\Omega)(^J\!A_\theta / \sum_{J=A}^{U} {}^J\!A_\theta)^{-1}$$

where $^{LJ}m_\Omega$ and $^{LJ}m_\theta$ are net migration rates (or probabilities) over time span Ω (the projection period, say 1981-1985) and θ (the base period, say 1976-1980) respectively; while $(^J\!A_\Omega / \Sigma {}^J\!A_\Omega)$ and $(^J\!A_\theta / \Sigma {}^J\!A_\theta)$ are measures of conditions (economic, social and/or political) in region J relative to the nation over time spans Ω and θ respectively.

On the other hand, when projections are being made for each demographic class d we may have a set of equations of the form:

$$^{LJ}m_{d\theta} = {}^{LJ}m_{d\theta} \, (^J\!A_{d\Omega} / \sum_{J=A}^{U} {}^J\!A_{d\Omega})(^J\!A_{d\theta} / \sum_{J=A}^{U} {}^J\!A_{d\theta})^{-1}$$

where $^{LJ}m_{d\Omega}$ is the net migration rate for demographic class d over time span Ω, $^J\!A_{d\Omega}$ is a measure of conditions in region J which are considered relevant by members of demographic class d when they make their migration decisions during Ω while $^{LJ}m_{d\theta}$ and $^J\!A_{d\theta}$ are defined in a parallel manner. For further discussion, see Shaw (1975), Hart (1975, 1975a), Greenwood (1975), Olsen, et al (1977), Isserman (1980, 1982) and Smith (1983, 295-301).

[27] See Carrothers (1956), Isard, et al (1960; Chapter 11) and Hua and Porcell (1979) for historical reviews of the use of the gravity concept in modelling spatial interaction. In addition, Hua and Porcell (1979) review a number of approaches for deriving gravity model formulations which do not rely on the above-mentioned analogy to Newtonian physics.

[28] Empirical estimates of migration-distance elasticities ranged from -0.50 to -1.50 in eleven studies reviewed by Shaw (1975). A number of hypotheses have been put forth attempting to explain the significance of these elasticities. For example, distance has been used as a proxy for transportation costs, psychic costs, and uncertainty (Greenwood, 1975). The correlation between transportation costs and distance is obvious. Psychic costs have been claimed to result from separation from family and friends; the greater the move, the lower the frequency of reunion and the higher the psychic cost (Schwartz, 1973). Finally, since information about a destination generally declines with distance, uncertainty increases -- and with this the expected value of the potential income stream at the destination declines (David, 1974).

[29] For a discussion of others, see Isard, et al (1960; Chapter 11), Lowry (1966) and Hua and Porcell (1979).

[30] For others, see Anselin and Isard (1979), Alonso (1977) and Ledent (1982) as well as the general literature on factors underlying regional outmigration patterns (eg. Shaw (1975), Greenwood (1975), Hart (1975, 1975a), Olsen, et al (1977) and Isserman (2980, 1982)).

[31] For others, see the general literature on the determinants of regional inmigration patterns (eg. Shaw (1975), Greenwood (1975), Olsen, et al (1977) and Isserman (1980, 1982)), as well as Anselin and Isard (1979), Alonso (1977) and Ledent (1982).

[32] Note that some of the variables used to measure the destination region mass may be identical to those used to measure the origin region mass. This is so because a given characteristic, say the regional unemployment level, may be considered a 'push-out' factor when high, and a 'pull-in' factor when low. And vice-versa. Others, such as a particular environmental factor, may be specific to a given destination (or origin) region.

[33] For alternative rationales for these exponents see, among others, Carrothers (1959), Isard, et al (1960), Wilson (1970), and Hua and Porcell (1979).

[34] Where the model is calibrated for one year flows, that is where $\theta=1$, then instead of an intermediate point type of calculation one may need to construct the model such that migration flows in year $t+\theta$ are related to conditions not only at year t but also at $t-\theta$, $t-2\theta$, and so on. This is so since migration tends to be a medium to long term decision, and the use of a one year lag may introduce unrealistic short-term fluctuations into migration projections.

[35] For an example of such an approach, see Kau and Sirmans (1979).

[36] However, the intuitive appeal of the cost efficiency principle (T. Smith, 1978, 1981) suggests that the negative exponential function may reflect human behaviour more closely than the power function. See Wilson (1970, p. 34) for discussion of the use of alternative distance functions when using the entropy maximising approach.

[37] See Hyman (1969), Batty and Mackie (1972), Evans (1971), Batty (1976) and Openshaw (1976) for a discussion of both the process of calibration and alternative measures of 'goodness of fit'.

[38] For a discussion of the population centroid and other measures of the central tendency of population distributions, see Shryock, Siegel, et al (1976), pp. 72-74.

[39] In effect, what one is doing resembles the use of a simple regression model when a multiple regression model is more appropriate.

[40] Alternatively, we should have:

$$^{J}O_{\Omega} = \prod_i [(^{J}X_{i\Omega})^{J}w_{i\Omega}] \text{ and } ^{L}D_{\Omega} = \prod_j [(^{L}Y_{j\Omega})^{L}v_{j\Omega}]$$

should a multiplicative rather than an additive form be used for the regional mass variables.

[41] Another approach would be to use a Saaty (1981) eigenvector procedure here and hence base the weights on so-called 'expert' opinion. However, since the above measures are not independent of one another (ie. weights may not be strictly additive) then this procedure may prove inadequate.

[42] See Gordon and Pitfield (1981) for an application of the use of an unconstrained gravity formulation in modelling age-sex disaggregated interregional migration flows in the United Kingdom.

[43] For example, for demographic classes including persons in the labour force (15-64) ages, a measure of regional labour market conditions (or a composite measure with high weights attached to labour market conditions) should be employed. While for persons aged 65 and over, an index of regional environmental quality may be more appropriate -- where such an index can be constructed. For persons aged under 15, migration rates are generally not independent of those of their parents -- so we may use the average migration rate projected for persons aged 15-64 (or some subset of this age group) as the measure of regional attractiveness for these persons.

[44] An earlier version of this section was included in Isard and Smith (1986).

[45]Alternatively, we may have a set of equations of the form

$$\ln(^{J\rightarrow}M_\theta/^JP_\theta) = \ln v_0 + \sum_{i=1}^{n} v_i \ln {}^JX_{\theta i} + \mu_2$$

where gross inmigration flows are felt to be a multiplicative rather than an additive function of the 'pull-in' factors $^JX_{\theta i}$ (i=1,...,n).

[46]Alternatively, we may have a set of equations of the form

$$\ln(^{\rightarrow L}M_\theta/^LP_q) = \ln u_0 + \sum_{j=1}^{m} u_i \ln {}^LY_{\theta i} + \mu_4$$

where gross outmigration flows are felt to be a multiplicative rather than an additive function of the 'push-out' factors $^LY_{\theta j}$ (j=1,...,m).

[47]See Ledent (1982) for an application of a model closely resembling the one proposed in this section. He employs the double-log form when estimating his regional in- and out-migration equations (5.3.17) and (5.3.18), respectively. Also in the spirit of the Alonso (1977) model he introduces some additional parameters α and β into the model, so that (5.3.19) becomes

$$^{JL}M_\theta = (^Ja)^{\alpha-1} (^Lb)^{\beta-1} \, ^JO_\theta \, ^LD_\theta \, f(^{JL}d").$$

[48]See Batty (1976) and Openshaw (1976) for a discussion of the procedures which may be employed in calibration of doubly constrained models of the form given in (5.3.19)-(5.3.21). Since we are dealing with net flows in this model, $\Sigma \, ^JO_\theta = \Sigma \, ^LD_\theta = 0$ can be regarded as another implicit constraint.

[49]Note that where more than one matrix of origin-destination specific migration flows are available, consideration may be given to fitting a doubly-constrained minimum information model of the form:

$$^{JL}M_\theta = g \, ^{JL}q_{\theta-1} \, ^Ja \, ^Lb \, ^JO_\theta \, ^LD_\theta \, f(^{JL}d")$$

where $^{JL}q_{\theta-1}$ is defined as the probablity that any individual in the entire system is a 'region J to region L' migrant during an observed base year period, and where $f(^{JL}d")$ is taken to be a negative exponential function of the form given in (5.3.14). Plane (1982) shows that the average relative percent error in replicating 1978-79 US interstate flows is only 7.4 percent if 1975-76 flows are used to fix the $^{JL}q_{\theta-1}$. A similar doubly constrained gravity model without the $^{JL}q_{\theta-1}$ terms had an average relative error of 45.8 percent.

[50] Alternatively, we may have

$$\ln(^{J\rightarrow}M_{d\theta}/^{J}\underline{P}_{d_\theta}) = \ln v_{d0} \sum_{i=1}^{n} v_{di} \ln {}^{J}X_{d\theta i} + \mu_7$$

and

$$\ln(^{\rightarrow L}M_{d\theta}/^{L}\underline{P}_{d\theta}) = \ln u_{d0} + \sum_{j+1}^{m} u_{di} \ln {}^{L}Y_{d\theta i} + \mu_8$$

where a multiplicative (rather than an additive) form is assumed for the relationship between migration flows and their corresponding socioeconomic determinants.

[51] See Stillwell (1978) for an application of the use of a doubly constrained gravity formulation in modelling age-sex disaggregated interregional flows in the United Kingdom. Since we are dealing with net flows in this model, $\Sigma\, {}^{J}O_{d\theta} - \Sigma\, {}^{L}D_{d\theta} = 0$ can be regarded as another implicit constraint.

[52] An earlier version of this section was included in Isard and Smith (1986).

[53] For example, these may be one of the official Census projections by the Australian Bureau of Statistics.

[54] If the inequality in (5.3.28) is very large, we may adjust g downwards by a region-specific factor a^L. However, another means of ensuring that projected regional unemployment levels fall within acceptable limits is described on page 173.

[55] Reasonable sounding arguments can be put forth in support of each of these approaches. On the one hand, it can be argued that a different set of factors determines outmigration than inmigration -- or that the time lags involved in response to a given factor differ between outmigration and inmigration. As a result, outmigration and inmigration should be modelled separately -- and net migration derived as an identity. On the other hand, it can be argued that the basic forces determining interregional migration flows can be captured through an examination of net flows -- with the perturbations causing cross-hauling etc. in gross flows reflected in the stochastic error term.

56A priori, one would expect the more disaggregated approach to yield superior results. This is so since (1) the migration patterns of different demographic classes often respond to a different set of factors (the classic case being the migration of the elderly) and (2) any given region may have a favourable score on a particular factor (say employment opportunities) at an aggregate level, but a highly unfavourable score on that factor for a number of demographic classes (say females and youth). However, the aggregate approach is not without its advantages. In particular, it avoids giving undue weights to random factors which may have dominated migration patterns of a particular demographic class -- with the result that it may generate more accurate projections of aggregate migration patterns.

57This is especially so when migration patterns within less developed countries are the objects of study.

58Discussion of the economic determinants of interregional migration is found in Shaw (1975), Greenwood (1975), Hart (1975, 1975a) and Isserman (1980, 1982). Popular candidates include regional wage rate differentials (neoclassical growth theory), differentials in the present value of expected lifetime earnings (human capital approach) and employment opportunities discounted by the probability of finding a job.

59Discussion of non-economic determinants of interregional migration is found in Cebula (1979), Roseman (1982) and Isserman (1980, 1982). Popular candidates include quality of life (pollution levels, crime levels, congestion, etc.), closeness of friends and relatives, climate (coastal locations, warmer locations), marital status changes, retirement, and school/university attendance. Unfortunately, many of these variables cannot be put in econometric models because of data availability problems. However, see Liu (1975) and Cebula (1979) for attempts to include 'quality of life' variables in their econometric models.

60There is also the problem of disciplinary bias which has long pervaded the study of migration phenomena. Economists, geographers, regional scientists, planners, demographers and sociologists have each been involved in migration modelling, and each group tends to emphasise those determinants which are most widely accepted in their own discipline.

61For example, it makes sense to include a variable 'probability of obtaining employment' in a model that postulates that migrants move in response to employment opportunities. However, implementing this approach requires a good empirical measure of this probability (see Todaro (1976), Fields (1976), and Levy and Wadycki (1974)).

62For example, consider the concept of 'alternative opportunties'. The likelihood of selecting a particular destination depends in part on conditions at alternative destinations. However, there is considerable disagreement in the literature on the definition of an appropriate set of alternatives. For example, some use all other regions; others all regions weighted by their distance from the origin region; and yet others all other regions no farther away from the origin than the given destination. See Levy and Wadycki (1974), Wadycki (1979), and Feder (1979, 1980) for relevant discussion.

[63] For example, does migration respond to present, past or expected future socio-economic conditions? If the migrant responds to expected future conditions, how far into the future does he look, and what 'discount rate' does he employ when comparing the immediate and distant future? If the migrant responds to past conditions, how far into the past does he look, and what 'discount rate' does he employ when comparing the immediate and distant past?

[64] For example, some analysts drop variables from these models when multicollinearity appears to be a major problem. Others make use of weighted least squares (or other data transformations) when heteroscedasticity appears to be a major problem. Some analysts impose constraints on their model to ensure that basic identities are not violated (eg. that the total net migration projected for the system is zero, and that net migration from any given region does not exceed its so called 'at risk' population). And so on.

[65] An earlier version of the remainder of this section was included in Isard and Smith (1986), but the latter was based on a preliminary version of this chapter. Note that it is recognized that some of the specifications given below might be considerably changed and improved after being confronted with empirical material.

[66] Note that equations (5.3.30) and (5.3.33) will be estimated for both gross and real wage income, and the formulation which generates the best results in terms of 'goodness of fit' employed in subsequent steps.

[67] However, since we have so few observations (even after pooling) we have to be careful that the addition of dummies does not use up too many degrees of freedom. There is also the question of whether these dummies should be employed in an additive or multiplicative fashion. Finally, consideration could be given to including a dummy variable for those regions (eg. Queensland) for which 'quality of life' (including climatic) factors are important additional variables influencing labour force migration.

[68] For a discussion of the nature of the problem of heteroscedasticity, see a standard econometrics text such as Johnston (1972, pp. 214-221) or Maddala (1977, pp. 259-274).

[69] Note that the possibility of employment demand being greater than labour supply is ignored in the above formulation. This is considered reasonable given the fact that such a circumstance will arise only very

rarely, and even then this will only be a temporary situation given the flexibility in both the wage rate and the laborforce participation rate variables incorporated in the labor market component of the multiregional econometric module.

[70] We have the additional problem that many of the same types of variables appear in both the in-migration and outmigration equations (5.3.32) and (5.3.33), respectively. As a result, it may be that one or both of these equations gives unexpected results (wrong signs, etc.) when confronted with actual data.

Note also that while some readers may imagine that interregional differences in the state of housing markets and/or quality of public infrastructure would play an important role in determining migration patterns, this is not so for Australian states -- where intra-regional differences in such variables are much more significant than any interregional differences.

[71] That is

$$(^{JL}M_{d\Omega}/^{J}\underline{P}_{d\Omega}) = \frac{[\sum_d {}^{JL}M_{d\Omega}/\sum_d s \, {}^{J}E_{d\Omega}]}{[\sum_d {}^{JL}M_{d\theta}/\sum_d s \, {}^{J}E_{d\theta}]} \, (^{JL}M_{d\theta}/^{J}\underline{P}_{d\theta})$$

(5.3.34)

and

$$(^{JL}M_{d''\Omega}/^{J}\underline{P}_{d''\Omega}) = \frac{[\sum_d {}^{JL}M_{d\Omega}/\sum_d s \, {}^{J}E_{d\Omega}]}{[\sum_d {}^{JL}M_{d\theta}/\sum_d s \, {}^{J}E_{d\theta}]} \, (^{JL}M_{d''\theta}/^{J}\underline{P}_{d''\theta})$$

(5.3.35)

where the subscripts Ω and θ refer to the projection period and the base year period respectively.

[72] That is,

$$[^{JL}M(65+)/^{J}\underline{P}(65+)] = {}^{JL}k \quad (J,L=A,...,U)$$

where ^{JL}k is determined on the basis of the latest population census results (or on the basis of time trends implied by the results of the latest population censuses). However, where region-specific changes occur in other relevant factors -- such as in tax conditions, public infrastructure provision or social welfare programs -- the analyst may make ad hoc adjustments in the corresponding ^{JL}k coefficients.

[73] One may ask why we have not proposed an exclusively econometric approach for modelling labour force migration flows. For example, one may hypothesise (as we originally did) that we would be better off with a behavioural equation for origin-specific, destination-specific migration flows rather than a gravity model formulation. That is, we might first estimate an equation for gross regional inmigration as in (5.3.32) above, and then set up an equation for origin-specific migration shares of the form:

$$(^{JL}M_d / ^{\Rightarrow L}M_d) = h_0 + h_1 \left[(^J WY_d /_D {}^J E_d) / (WY_d /_D E_d) \right]$$

$$+ h_2 \frac{\Delta(^J WY_d /_D {}^J E_d) / (^J WY_d /_D {}^J E_d)}{\Delta(WY_d /_D E_d) / (WY_d /_D E_d)}$$

$$+ h_3 \left[(_D {}^J E_d /_S {}^J E_d) /_D E_d /_S E_d) \right] + \mu_5 \qquad (5.3.36)$$

the levels of gross regional outmigration could then be derived via a set of identities of the form:

$$^{J \Rightarrow} M_d = \sum_L \left[(^{JL}M_d / ^{\Rightarrow L}M_d) \cdot {}^{\Rightarrow L}M_d \right].$$

However, a number of problems arise when we try to estimate equation (5.3.36). Firstly, we have the problems discussed earlier with respect to the appropriate timing of observations and inclusion of 'migration gains' in related demographic classes. Secondly, we have some additional problems as well. In particular, the conventional ordinary least squares regression technique cannot be employed, since it will not guarantee that the migration shares $(^{JL}M_{d\Omega} / ^{\Rightarrow L}M_{d\Omega})$ fall between 0 and 1. One useful transformation which can be used under the circumstances is that underlying the mulinomial logit model, in which case the dependent variable in (5.3.36) becomes

$$\ln[(^{JL}M_d / ^{\Rightarrow L}M_d) / \{1 - (^{JL}M_d / ^{\Rightarrow L}M_d)\}]$$

This transformation guarantees that the shares each fall between 0 and 1. However, it does not guarantee that they sum to 1. Further, since we have an equation (5.3.32) for estimating the total migration into region

L, we must drop one of the share equations from the set given by (5.3.36) -- with this region's share being determined as a residual. However, in general, the results can be expected to vary depending on which share is treated as a residual. This is not a highly satisfactory state of affairs; however it may be argued that after some iterations we may be able to identify a share which can be dropped and projected satisfactorily as a residual. But this is not the case for a situation where time series and cross section data has to be pooled in order to yield reasonable parameter estimates. It is difficult to interpret what one is really estimating when one attempts to calibrate a multinomial logit model using pooled time series and cross-sectional data -- and hence we feel more comfortable dismissing this type of approach until a reasonable time series of interregional migration data becomes available.

[74] This is done by the use of a set of 'reasonable' population to final demand coefficients for each final demand category and region. These coefficients can relate to either total population or be broken down by demographic class, depending on the type of final demand being projected.

[75] These coefficients can relate to the total labour force or be broken down by demographic class, depending on the level of disaggregation required for operation of the migration component of the demographic model.

CHAPTER 6
A MULTIREGIONAL ECONOMETRIC MODULE AS A COMPONENT OF AN INTEGRATED MULTIREGIONAL MODEL

6.1 Introduction

This chapter is concerned with the development of a multiregional econometric module which is capable of forming a component of an integrated multiregional model. In Section 6.2 we outline the structure of such a module capable of implementation in the Australian context. For each region it includes equations for determining household income by type, consumption expenditure by type, government expenditure by type by level, private investment expenditure by sector, and employment demand, labour supply and wage rates by demographic class. As discussed in Section 6.3, the projections derived from operating an econometric module of this form can be used to revise the final demand vector employed when operating the interregional input-output module. In Section 6.4, we outline the steps involved in integrating multiregional input-output, demographic and econometric modules. The operation of a more fully integrated multiregional model comprising: (a) an interregional input-output module, (b) a comparative cost-industrial complex module, (c) an interregional programming module, (d) a multiregional demographic module, and (e) a multiregional econometric module is discussed in Section 6.5. Finally, some concluding remarks are made in Section 6.6.

6.2 Outline of the Multiregional Econometric Module

The multiregional econometric module to be presented in this section contains four sets of equations for each region: (a) a household income and consumption expenditure submodel, (b) a labour market submodel, (c) a private investment expenditure submodel, and (d) a government expenditure submodel. We now discuss each submodel in turn.

6.2.1 Household Income and Consumption Expenditure Submodel

The household income and consumption expenditure submodel includes equations for determining seven components of household income, and four types of consumption expenditure. Figure 6.1 depicts the general relationships among the variables included in this submodel.[1]

a. Household Income Equations[2]

Total household income by region of residence is derived from two sources: wage and salary income (earnings) and non-wage income.

Wage and Salary Income

The first variable, wage and salary income WY, is modelled as an identity. That is it is obtained for each region J (J=A,...,U) by multiplying demographic class specific wage rates for each industry i (i=1,...,n) by their respective employment demand levels, and then summing over all industries and demographic classes d (d=1,...,d). That is:[3]

$$^{J}WY_{idt} = {}^{J}WR_{idt} \cdot {}_{D}^{J}E_{idt} \qquad (6.2.1)$$

$$^{J}WY_{t} = \sum_{i} \sum_{d} {}^{J}WY_{idt} \qquad (6.2.2)$$

where $^{J}WY_{idt}$ = wages and salaries paid to members of demographic class d by sector i in region J,

$_{D}^{J}E_{idt}$ = employment demand (requirements) for members of d by sector i, region J (measured in manhours), and

$^{J}WR_{idt}$ = wage rate (per manhour) for members of demographic class d in sector i, region J.

ECONOMETRIC MODULE

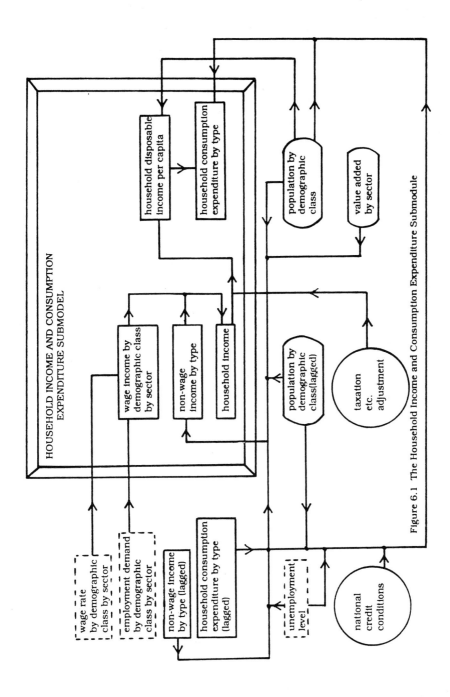

Figure 6.1 The Household Income and Consumption Expenditure Submodule

Non-Wage Income

There are six components of non-wage income ^{J}NWY estimated by this submodel: supplements to wages and salaries, income from farm unincorporated enterprises, income from non-farm unincorporated enterprises, income from dwellings, transfers from government, and other non-wage income. That is, for each region J we have an identity of the form:

$$^J\text{NWY}_t = \sum_{k=1}^{6} {}^J\text{NWY}_{kt} \qquad (6.2.3)$$

where ^{J}NWY$_{kt}$ = non-wage income accruing to residents of region J from source k.

Because the nature of the six types of income vary substantially, each is estimated using a different specificiation:

(1) <u>Supplements to wages and salaries</u> ^{J}NWY$_{1t}$ are calculated for each region J by assuming that they comprise a fixed proportion $^J\delta_i$ of the wage and salary income earned in each industry i.[4] That is:

$$^J\text{NWY}_{1t} = \sum_i [{}^J\delta_i \, {}^J\text{WY}_{it}] \qquad (6.2.4)$$

(2) <u>Incomes from farm and non-farm unincorporated enterprises</u> are calculated respectively as:

$$^J\text{NWY}_{2t} = \alpha_0 + \alpha_1 \, ({}^J\text{VA}_{i't} - {}^J\text{WY}_{i't}) + \alpha_2 \, \text{TIME} + \mu_{1t} \qquad (6.2.5.)$$

and

$$^J\text{NWY}_{3t} = \beta_0 + \beta_1 \sum_i ({}^J\text{VA}_{it} - {}^J\text{WY}_{it}) + \beta_2 \, \text{TIME} + \mu_{2t} \qquad (6.2.6)$$

where $^J VA_{i't}$ = value added by sector i' (= agriculture) in region J;

$^J WY_{i't}$ = wage and salary income paid out by sector i' in region J;

$^J VA_{it}$ = value added by sector i (i= manufacturing or other non-manufacturing) in region J;

$^J WY_{it}$ = wage and salary income paid out by sector i in region J;

TIME = a trend variable included to reflect changes in the proportion of gross operating surplus accruing to unincorporated enterprises in the given industy.

(3) The <u>income from dwellings</u> specification, shown in equation (6.2.7), includes three basic types of expanatory variables: a demand for housing variable -- represented by change in population ($^J P_t - {}^J P_{t-1}$) in region J, an ability to pay for housing variable -- represented by wage and salary income in region J, and a lagged dependent variable.[5] That is:

$$^J NWY_{4t} = \gamma_0 + \gamma_1 ({}^J P_t - {}^J P_{t-1})$$
$$+ \gamma_2 {}^J WY_t + \gamma_3 {}^J NWY_{4t-1} + \mu_{3t} \qquad (6.2.7)$$

(4) <u>Transfers from government</u> $^J NWY_{5t}$ are estimated using the following equation:

$$^J NWY_{5t} = \emptyset_0 + \emptyset_1 {}^J WY_t + \emptyset_2 {}^J P_{(>60)t}$$
$$+ \emptyset_3 {}^J UN_t + \emptyset_4 P^J_{(\leq 15)t} + \mu_{4t} \qquad (6.2.8)$$

where $^J WY_t$ = wage and salary income accruing to residents of region J is included since (generally speaking) the lower the value of this variable, the greater the reliance on transfers from government,

$^J P_{(>60)t}$ = population aged over 60 years is included because old age pensions comprise a significant proportion of $^J NWY_{5t}$.

$^J UN_t$ = the unemployment level in region J is included because unemployment benefits comprise a significant proportion of $^J NWY_{5t}$, and

$^J P_{(\leq 15)t}$ = population aged 15 years or younger is included since family allowances (and other forms of child support) comprise a significant proportion of $^J NWY_{5t}$.

(5) Finally, <u>other non-wage income</u>[6] is estimated as:

$$^J NWY_{6t} = \theta_1 + \theta_2 \left[^J WY_t + \sum_{k=1}^{5} {^J NWY_{kt}} \right] + \theta_3 \text{ TIME} + \mu_{5t}$$

(6.2.9)

Thus it is related to other components of household income (the second term on the RHS), with a trend factor (the third term on the RHS) included to capture the effect of any shifts over time in the proportion of total household income accounted for by the non-wage income component.

Household Disposable Income

Household disposable income $^J HDY$ can then be estimated for each region J by adding the wage and salary income derived in equation (6.2.2) to the total non-wage income derived in equation (6.2.9), and then adjusting for taxation payments and other forms of non-disposable income.[7] That is, we make use of the following identities:

$$^J HY_t = {^J WY_t} + {^J NWY_t}$$

(6.2.10)

and

$$^J HDY_t = (1 - {^J \psi_t}) {^J HY_t}$$

(6.2.11)

where $^J HY_t$ = household income accruing to residents of region J;

$^J\psi_t$ = proportion of household income accounted for by taxation payments and other forms of non-disposable income.[8]

The level of household disposable income per capita in region J, $^JHDY_t$, is then given by:

$$^J\underline{HDY}_t = {^JHDY_t}/{^J\underline{P}_t} \qquad (6.2.12)$$

where $^J\underline{P}_t$ = population level in region J.

b. Household Consumption Expenditure Equations[9]

There are four types of household consumption expenditure ^JHCE estimated by this submodel: food, drink and tobacco expenditures, other non-durable expenditures, durables (including motor vehicles) expenditures, and dwelling rent. That is, for each region J we have an identity of the form:

$$^JHCE_t = \sum_{q=1}^{4} {}^JHCE_{qt} \qquad (6.2.13)$$

where $^JHCE_{qt}$ = household consumption expenditure of type q made by residents of region J.

Because the nature of these expenditure types vary substantially, we make use of a number of different specifications.

(1) <u>Food, drink and tobacco expenditures</u> $^JHCE_{1t}$ and <u>other non-durable expenditures</u> $^JHCE_{2t}$ are each estimated in per capita terms using an equation of the same form. In particular, for q=1,2 we have:

$$(^JHCE_{qt}/{^J\underline{P}_t}) = \delta_0 + \delta_1 {}^J\underline{HDY}_t + \delta_2 \text{ TIME} + \mu_{6t} \qquad (6.2.14)$$

Thus these expenditures are related to household disposable income per capita (the second term on the RHS), with a time factor (the third term on the RHS) included to capture the effects of any shift over time in the proportion of household disposable income spent on the given category of consumption.

(2) The equation for <u>expenditures on durables (including motor vehicles)</u> $^{J}HCE_{3t}$ includes more explanatory variables, yet is also estimated in per capita terms. That is:

$$(^{J}HCE_{3t}/^{J}P_{t}) = \lambda_0 + \lambda_1 {}^{J}HDY_t + \lambda_2 {}^{J}UN_t$$
$$+ \lambda_3 \$_t + \lambda_4 (^{J}HCE_{3t-1}/^{J}P_{t-1}) + \mu_{7t} \qquad (6.2.15)$$

where $^{J}HDY_t$ = household disposable income per capita in region J is included since (generally speaking) the higher the value of this variable, the larger the expenditure on durable consumption items;

$^{J}UN_t$ = unemployment level in region J is included since (generally speaking) the higher the value of this variable, the lower the expenditure on durable consumption items;

$\$_t$ = an indicator of national credit conditions (for example, the prevailing interest rate on bank overdrafts) is included since a large proportion of durable goods expenditure is carried out using various forms of consumer credit; and

$^{J}HCE_{3t-1}/^{J}P_{t-1}$ = the lagged dependent variable is included to reflect a partial adjustment mechanism.[10]

(3) The <u>dwelling rent payments</u> $^{J}HCE_{4t}$ specification, shown in equation (6.2.16), includes three basic types of explanatory variables: a demand for housing variable -- represented by change in population ($^{J}P_t$ - $^{J}P_{t-1}$) in region J, an ability to pay for housing variable -- represented by the level of household disposable income, and a lagged dependent variable.[11] That is:

$$^JHCE_{4t} = \infty_0 + \infty_1 (^JP_t - ^JP_{t-1})$$
$$+ \infty_2 {}^JHDY_t + \infty_3 {}^JHCE_{4t-1} + \mu_{8t} \qquad (6.2.16)$$

6.2.2 Labour Market Submodel

The labour market submodel includes equations for determining wage rates, employment demand and labour supply by demographic class. Figure 6.2 depicts the general relationships among the variables included in this submodel.

a. Employment Demand Equations

In this module we continue to assume that the demand for employees in demographic class d (d=1,...d) can be determined for each sector i (i=1,...,n) and region J (J=A,...,U) via the use of a set of fixed coefficients.[12] That is:

$$_D{}^JE_{it} = {}^Jw_{it} \, {}^JX_{it} \qquad (6.2.17)$$

and

$$_D{}^JE_{idt} = {}^Jw_{idt} \, _D{}^JE_{it} \qquad (6.2.18)$$

where $^Jw_{it}$ = number of manhours of labour required per unit output of sector i in region J,

$^Jw_{idt}$ = proportion of sector i's labour requirements which is met by members of demographic class d.

An alternative approach which could be adopted for manufacturing and mining sectors would be the use of a set of equations of the form:[13]

$$_D{}^JE_{idt} = a_0 + a_1 \, {}^JVA_{it} + a_2 \, (^JWR_{idt}/{}^JWR_{it})$$
$$+ a_3 \, (UCC_t/{}^JWR_{it}) + \mu_{1t} \qquad (6.2.19)$$

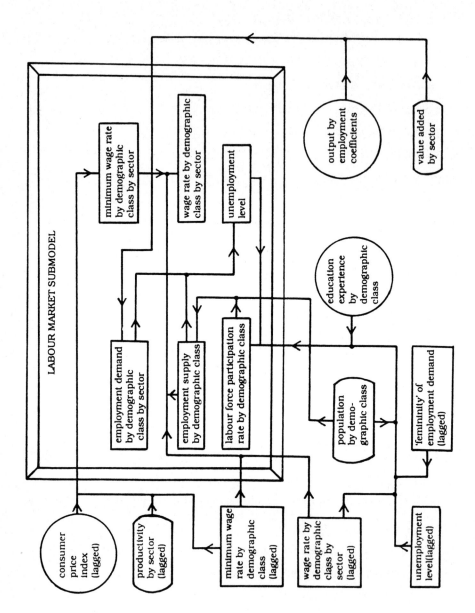

Figure 6.2 The Labour Market Submodule

where $^J VA_{it}$ = value added by sector i in region J,

$^J WR_{idt}/^J WR_{it}$ = wage rate of persons in demographic class d employed in sector i relative to the average wage rate of persons employed in that sector,

$UCC_t/^J WR_{it}$ = user cost of capital relative to the average wage rate of persons employed in sector i, J.

Such a formulation allows for (1) changes in output to employment ratios as output levels increase, and (2) changes in capital to labour ratios as wage rates increase relative to the user cost of capital.[14] However, this in turn suggests the need for input-output coefficient modifications, and hence would complicate considerably the linkage of the input-output and econometric modules. As a result we prefer to postpone consideration of this important aspect of reality until Chapter 7, where we discuss the inclusion of a factor demand module as a component of an integrated multiregional model.

b. Labour Supply Equations

For each region J, we obtain the supply of labour $_S{}^J E_d$ in each demographic class d (d=1,...,d) via the use of a set of identities of the following form:

$$_S{}^J E_{dt} = (^J LF_{dt}/^J \underline{P}_{dt}) \cdot ^J \underline{P}_{dt} \cdot ^J H_{dt} \qquad (6.2.20)$$

where $_S{}^J E_{dt}$ = number of manhours of labour supplied by members of demographic class d in region J,

$(^J LF_{dt}/^J \underline{P}_{dt})$ = labour force participation rate for members of demographic class d in region J,

$^J LF_{dt}$ = number of persons in demographic class d who are members of the labour force in region J, and

$^J H_{dt}$ = average number of manhours of labour supplied by members of demographic class d in region J.[15]

The next step we take is to recognize that labour force participation rates are not fixed, but are influenced by labour market conditions. In doing so, we recognize differences between demographic classes by using different functional forms for modelling participation rates. In particular, we proceed as follows.[16]

Males aged 25-64

For demographic classes which fall in this age-sex range we employ an equation of the form:

$$(^J LF_{dt}/^J P_{dt}) = b_0 + b_1\, ^J WR_{dt-1} + b_2\, [(^S\!^J E_{d,t-1} - ^D\!^J E_{d,t-1})/^S\!^J E_{d,t-1}] + \mu_{2t} \qquad (6.2.21)$$

where lagged wage rates $^J WR_{d,t-1}$ are included as a proxy for gains from employment, and lagged unemployment rates[17] $(^S\!^J E_{d,t-1} - ^D\!^J E_{d,t-1})/^S\!^J E_{d,t-1}$ is added as a proxy for the probability of finding employment upon entering the labour force.

Females aged 25-64

The specification employed for demographic classes falling in this age-sex range, shown in equation (6.2.22), also includes lagged wage rates and lagged unemployment rates as explanatory variables. However in addition we have:

(1) the median years of school completed $^J EDU_{dt}$ by members of demographic class d in region J -- generally speaking the higher the value of this variable, the higher the female workforce participation rate,

(2) the number of dependents (persons aged under 15 years of age) per capita $(^J \underline{P}(\leq 15)_t / ^J \underline{P}_t)$ in region J -- generally speaking the higher the value of this variable, the lower the female workforce participation rate, and

(3) an index of 'femininity' of employment demand in region J, $(_D{}^JE_{F,t-1}/_D{}^JE_{t-1})$[18] since females may be more willing to enter the workforce in those regions which have established a 'reputation' for employing a high proportion of females. That is:

$$(^JLF_{dt}/^JP_{dt}) = c_0 + c_1\ ^JWR_{d,t-1}$$
$$+ c_2\ [(_S{}^JE_{d,t-1} - _D{}^JE_{d,t-1})/_S{}^JE_{d,t-1}] + c_3\ ^JEDU_{dt}$$
$$+ c_4\ [^JP_{(\le 15)t}/^JP_t] + c_5\ (_D{}^JE_{F,t-1}/_D{}^JE_{t-1}) + \mu_{3t} \qquad (6.2.22)$$

Persons aged 15-24

For persons aged 15-24, labour force participation rates are affected by the extent to which members of the given demographic class engage in full time educational activities. As a result, we employ a set of equations of the form:

$$(^JLF_{dt}/^JP_{dt}) = e_0 + e_1\ ^JWR_{d,t-1}$$
$$+ e_2\ [(_S{}^JE_{d,t-1} - _D{}^JE_{d,t-1})/_S{}^JE_{d,t-1}]$$
$$+ e_3\ (^JSCH_{dt}/^JP_{dt}) + \mu_{4t} \qquad (6.2.23)$$

where $(^JSCH_{dt}/^JP_{dt})$ = proportion of the population in demographic class d who are full-time students in region J.

Persons aged 65 and over

Workforce participation by persons aged 65 and over is regarded as either purely voluntary (workalcoholics and the like) or due to economic necessity. In either case, we assume that these workers represent a constant proportion of each region's population in their corresponding demographic classes. That is:

$$(^JLF_{dt}/^JP_{dt}) = {}^J\hat{O}_d \qquad (6.2.24)$$

where the relevant $^J\hat{O}_d$ are determined on the basis of past data.

c. Wage Rate Equations[19]

As in Section 6.2.1 the wage rate variables $^JWR_{idt}$ dealt with in this submodel are demographic class (d=1,...,d) and industrial sector (i=1,...,n) specific. Due to data limitations, each region's economy is divided into only four sectors: manufacturing, agricultural, defence, and other non-manufacturing.[20] We then proceed as follows:

Manufacturing, Other Non-Manufacturing Sectors

While wage rates in many countries are determined in a market, in Australia wage rates are composed of two major components:

(1) a basic wage component, which is determined in the arbitration courts when national and state wage decisions are handed down, and

(2) an above award wage component, which is determined for each region on the basis of its local labour market conditions.

The specification of the wage rate equations for these two classes of sectors reflects this wage formation process. To be more specific, we have equations of the general form specified below. For the nation:

$$\frac{(BW_{idt} - BW_{idt-1})}{BW_{idt-1}} = f_0 + f_1 \, \Delta CPI + f_2 \, \Delta(VA_i/_D E_i) + \mu_{5t}$$

$$\text{(t-2)to(t-1)} \qquad \text{(t-2)to(t-1)}$$

(6.2.25)

and for each region J:

$$\frac{(^JBW_{idt} - {^J}BW_{idt-1})}{^JBW_{idt-1}} = {^J}g_0 + {^J}g_1 \frac{(BW_{idt} - BW_{idt-1})}{BW_{idt-1}} + {^J}g_2 \, TIME + {^J}\mu_{6t}$$

(6.2.26)

where BW_{idt} = basic wage rate (per manhour) for demographic class d when employed by sector i;

ΔCPI = change in consumer price index (lagged);
(t-2) to (t-1)

$\Delta(VA_i/{}_DE_i)$ = change in labour productivity in sector (lagged);[21]
(t-2) to (t-1)

TIME = a time trend included to reflect the increasing proportion of wage decisions covered by national rather than state awards, and hence a convergence of basic wage rates towards the national average in each state.[21]

Thus, we have national basic wage rate changes determined on the basis of cost of living and productivity adjustments; while regional basic wage rate changes are determined as a function of their national counterparts.

The relationship between total wage rates and basic wage rates in any sector is then determined for each region as a function of local labour market conditions via a partial adjustment model. That is:

$$(^JWR_{idt}/{}^JBW_{idt}) = h_0 + h_1 ({}_D{}^JE_{dt}/{}_S{}^JE_{dt})$$
$$+ h_2 (^JWR_{idt-1}/{}^JBW_{idt-1}) + \mu_{7t} \qquad (6.2.27)$$

Thus the ratio between labour demand ${}_D{}^JE_{dt}$ and labour supply ${}_S{}^JE_{dt}$ is taken as the appropriate measure of labour market 'tightness' for members of demographic class d.[22] We use a partial adjustment model to reflect inertia and 'stickiness' in the labour market -- that is, while wage rates respond to labour market 'tightness' in a given region they do so with a lag.

Agricultural Sectors

Wage rate changes in agricultural sectors are assumed to follow those occurring in non-agricultural sectors. However, since local labour market conditions can act to modify the nature of this leader-follower pattern[23] we have a set of equations of the form:

$$^J WR_{it} = k_0 + k_1\, ^J WR_{jt}$$
$$+ k_2\, [_D{}^J E_{it}/(_S{}^J E_t - {}_D{}^J E_{jt})] + \mu_{8t} \qquad (6.2.28)$$

where $^J WR_{it}$ = wage rate (per manhour) in agricultural sectors i in region J;

$^J WR_{jt}$ = wage rate (per manhour) in non-agricultural sectors j in region J;

$(_S{}^J E_t - {}_D{}^J E_{jt})$ = 'excess' labour supply in region J after deducting non-agricultural sector demands;

$_D{}^J E_{it}$ = employment demand (in manhours) by agricultural sectors in region J.

Defence Sector

Wage rates in the defence sector are assumed to be identical in all regions, and to change in proportion to the national average increase excluding this sector.[24] However, this assumption can be modified should information become available on:

(1) rates of change in defence sector wage rates relative to their civilian counterparts, and/or

(2) regional differences in defence sector wage rates.

6.2.3 Private Investment Expenditure Submodel

The private investment expenditure submodel includes equations for determining gross fixed capital expenditure $^J KPR_{it}$ by 2-digit ASIC industrial sectors.[25] Figure 6.3 depicts the general relationships among the variables included in this submodel.

a. Agricultural, Manufacturing and Mining Sectors

The equation used for determining investment levels in these sectors is of the form:

ECONOMETRIC MODULE

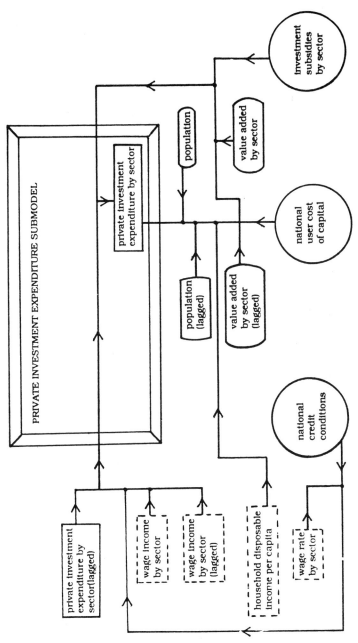

Figure 6.3 The Private Investment Expenditure Submodule

$$^J\text{KPR}_{it} = q_0 + q_1 \ [(^J\text{VA}_{it} - {}^J\text{WY}_{it}) - (^J\text{VA}_{it-1} - {}^J\text{WY}_{it-1})]$$
$$+ q_2 \ (^J\text{WR}_{it}/\text{UCC}_{it}) + q_3 \ (^J\text{KS}_{it}/\text{KS}_{it})$$
$$+ q_4 \ {}^J\text{KPR}_{i,t-1} + \mu_{1t} \tag{6.2.30}$$

That is, for any given region J investment levels in each of these sectors is taken to be a function of

(1) changes in the difference between value added $^J\text{VA}_i$ and the wage bill $^J\text{WY}_i$. This variable is included to capture the influence of profit level changes on the expectations of firms in the given sector, as well as on the ability of these firms to finance new investment;

(2) the ratio between wage rates $^J\text{WR}_{it}$ and the national user cost of capital UCC_{it}. In general we would expect an increase in this ratio to motivate firms in the given sector to attempt to substitute capital for labour, and hence invest in labour-saving machinery and equipment;

(3) the subsidies available to firms in the given sector to help finance investment in region J relative to those available should they choose to invest elsewhere in the nation -- that is $(^J\text{KS}_{it}/\text{KS}_{it})$;[26] and

(4) the lagged dependent variable $^J\text{KPR}_{i,t-1}$. This variable is included to capture the effects of lags in the adjustment of investment expenditures to changed economic opportunities.

b. Residential Construction Sector

The specification for investment (capital) expenditure on residential construction, shown in equation (6.2.31) includes four basic types of explanatory variables:

(1) a demand for housing variable -- represented by change in population in region J;

(2) an ability to pay for housing variable -- represented by the level of household disposable income in region J;

(3) an indicator of national credit conditions (for example, the prevailing interest rate on bank overdrafts) -- since residential

construction expenditure is often highly sensitive to these conditions; and

(4) a lagged dependent variable -- reflecting a partial adjustment of construction expenditure to changes in the effective demand for housing in the current period.

That is, we have[27]

$$^J KPR_{it} = r_0 + r_1 (^J \underline{P}_t - ^J \underline{P}_{t-1}) + r_2 \,^J HDY_t + r_3 \$_t$$
$$+ r_4 \,^J KPR_{i,t-1} + \mu_{2t} \qquad (6.2.31)$$

c. Other Sectors

Region-specific investment expenditure data is not available for other sectors. As a result, we assume that the national investment KPR_{it} carried out by each of these sectors is allocated among regions in proportion to the value added by that sector in the given regions. To be more specific, we make use of a set of equations of the form:

$$^J KPR_{it} = (^J VA_{it}/ VA_{it}) KPR_{it} \qquad (6.2.32)$$

The national investment level KPR_{it} in any given sector i can be obtained from a national econometric module, or alternatively from the use of an equation of a form similar to that given by (6.2.30) above.

6.2.4 Government Expenditure Submodel

The government expenditure submodel includes equations for determining eight categories of final consumption expenditure, and seven categories of gross fixed capital expenditure for each level of government (federal, state and local). Figure 6.4 depicts the general relationships among the variables included in this submodel.

226 CHAPTER 6

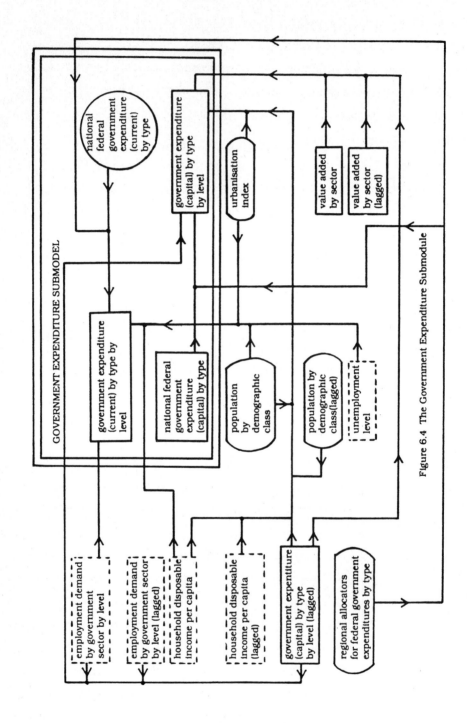

Figure 6.4 The Government Expenditure Submodule

a. Federal Government Expenditure Equations[28]

Let $^J GCONS_{gpt}$ be the level of current (consumption) expenditure by government level g (g =1,2,3) on function p (p=1,...,8) in region J, and $^J GINV_{gqt}$ be the level of capital (investment) expenditure by government level g on function q (q=1,...,7) in region J.[29] We then take national expenditures by the federal government (g=1) to be exogenous, and allocate these expenditures among regions via the use of an appropriate set of constants,[30] That is:

$$^J GCONS_{1pt} = {}^J\psi_{pt}\, GCONS_{1pt}$$

$$^J GINV_{1qt} = {}^J\Omega_{qt}\, GINV_{1qt}$$

(6.2.33)

b. State and Local Government Current Expenditure Equations[31]

Because the functions (p=1,...,8) performed by government authorities vary substantially, the level of expenditure on each is estimated using a different specification.

Public Administration (p=1)

Current expenditures on performing the administrative functions of government in each region J are taken to depend on the labour requirements $_D{}^J E_{gt}$ (measured in manhours) of government authorities operating in that region. That is, for each level of government g (=2,3) we have:

$$^J GCONS_{g1t} = \alpha_0 + \alpha_1 {}_D{}^J E_{gt} + \alpha_2\, TIME + \mu_{1t} \qquad (6.2.34)$$

where the trend factor (the third term on the RHS) is included to capture the effects of changes over time in the 'labour-intensiveness' of the administrative tasks performed by g,J.

Education (p=2)

For each level of government g (=2,3) in region J we calculate the current expenditures associated with the provision of educational services as:

$$^JGCONS_{g2t} = \beta_0 + \beta_1 \,^JSCH_t + \beta_2\, TIME + \mu_{2t} \qquad (6.2.35)$$

where $^JSCH_t$ = the number of full-time students resident in region J; and
TIME = a trend factor included to capture the effects of changes over time in per capita expenditures on the education function.

Health, Recreation and Culture (p=3,6)

Current expenditures on performing these two functions of government in each region J are taken to depend on population levels ^JP_t and the level of household disposable income per capita $^JHDY_t$. That is for each level of government g (=2,3) we have:

$$^JGCONS_{gpt} = \gamma_0 + \gamma_1 \,^JP_t + \gamma_2 \,^JHDY_t + \gamma_3\, TIME + \mu_{3t} \qquad (6.2.36)$$

where the trend factor (the fourth term on the RHS) is once again included to capture the effects of changes over time in per capita expenditures on function p.

Social Security and Welfare (p=4)

The specification employed for this function of government (given in equation (6.2.37)) also includes population levels, the level of household disposable income per capita and a trend factor as explanatory variables. However, in addition it includes (1) the rate of unemployment $(_S{}^JE_t - _D{}^JE_t)/_S{}^JE_t$, and (2) the proportion of the population over 65 years of age $(^JP_{(\geq 65)t}/^JP_t)$ in the given region -- since the unemployed and the elderly comprise two important categories of welfare recipients. That is:

$$^J GCONS_{g4t} = \delta_0 + \delta_1\, ^J\underline{P}_t + \delta_2\, ^J HDY_t + \delta_3\, [(^S{}^J E_t - {}^D{}^J E_t)/^S{}^J E_t]$$
$$+ \delta_4\, (^J\underline{P}_{(\geq 65)t}/^J\underline{P}_t) + \delta_5\, TIME + \mu_{4t} \qquad (6.2.37)$$

Housing and Community Amenities (p=5)

For each level of government g (=2,3) in region J we calculate the level of current expenditures associated with the provision of housing and community amenities as:

$$^J GCONS_{g5t} = \varepsilon_0 + \varepsilon_1\, ^J\underline{P}_t + \varepsilon_2\, ^J HDY_t$$
$$+ \varepsilon_3\, ^J URB_t + \varepsilon_4\, TIME + \mu_{5t} \qquad (6.2.38)$$

where $^J URB_t$ = an urbanization index (for example, the proportion of region J's population which is resident in metropolitan areas). This variable is included since the level of urbanization effects the cost of providing a given level of social/economic infrastructure to a region's population.

Economic Services (p=7)

We use a different equation for each sector i (i=1,...,5) which receives economic services from government authorities.[32] However, the same general form is employed for each sector and level of government. That is:

$$^J_i GCONS_{g7t} = \lambda_0 + \lambda_1\, ^J VA_{it} + \lambda_2\, TIME + \mu_{6t} \qquad (6.2.39)$$

where $^J VA_{it}$ = the value added in region J by producers in sector i, and

TIME = a trend factor included to capture the effects of changes over time in the level of government assistance given to sector i.

Defence (p=8)

Since the responsibility for national defence rests soley with the federal government, expenditures on this function are equal to zero for state and local government in each region J.

c. State and Local Government Capital Expenditure Equations

For each function (p=1,...,8), the equations used to project regional gross fixed capital expenditures by state and local governments are similar in form to the corresponding ones described in the previous section. In particular, we make the following changes:

(1) For expenditures on public administration, equation (6.2.34) is replaced by

$$^J GINV_{g1t} = a_0 + a_1 (_D{}^J E_{gt} + _D{}^J E_{gt-1})$$
$$+ a_2 {}^J GINV_{g1t-1} + \mu_{1t} \qquad (6.2.40)$$

(2) For expenditures on education, equation (6.2.35) is replaced by

$$^J GINV_{g2t} = b_0 + b_1 (^J SCH_t - {}^J SCH_{t-1})$$
$$+ b_2 {}^J GINV_{g2t-1} + \mu_{2t} \qquad (6.2.41)$$

(3) For expenditures on health and recreation and culture, equation (6.2.36) is replaced by

$$^J GINV_{gpt} = c_0 + c_1 (^J \underline{P}_t - {}^J \underline{P}_{t-1}) + c_2 {}^J HDY_t$$
$$+ c_3 {}^J GINV_{gpt-1} + \mu_{3t} \qquad (6.2.42)$$

(4) For expenditures on social security and welfare, equation (6.3.37) is replaced by

$$^J GINV_{g4t} = e_0 + e_1 (^J P_t - {^J P_{t-1}}) + e_2 {^J HDY_t}$$
$$+ e_3 {^J GINV_{g4t-1}} + \mu_{4t} \qquad (6.2.43)$$

(5) For expenditures on housing and community amenities, equation (6.2.38) is replaced by

$$^J GINV_{g5t} = f_0 + f_1 (^J P_t - {^J P_{t-1}}) + f_2 {^J HDY_t}$$
$$+ f_3 {^J URB_t} + f_4 {^J GINV_{g5t-1}} + \mu_{5t} \qquad (6.2.44)$$

(6) For expenditures on the provision of economic services to industry, equation (6.2.39) is replaced by

$$^J_i GINV_{g7t} = h_0 + h_1 (^J VA_{it} - {^J VA_{it-1}})$$
$$+ h_2 {^J_i GINV_{g7t-1}} + \mu_{6t} \qquad (6.2.45)$$

Thus in general we assume that capital expenditure by each level of government responds to changes in the demand for the particular service in question, but that it does so with a lag.

6.3 Integration of Multiregional Input-Output and Econometric Modules

The use of an interregional input-output model for estimating the level of economic activity at some point of time in the future -- say year t+θ -- requires that projections of the level of final demand by type (household consumption, private investment, government expenditure, etc.) be available for year t+θ for each region. One means via which the latter projections can be made is via the use of a multiregional econometric module of the form discussed in Section 6.2. In this section, we present the basic steps required to achieve a meaningful integration of these two types of models -- taking one econometric submodel up at a time.[33]

6.3.1 Linkage with Household Income and Consumption Expenditure Submodel

Step 1: *Derivation of Initial Final Demand Projections*

The first step involves the projection of an initial final demand vector -- including household consumption expenditures, private investment expenditures and government expenditures -- for year $t+\theta$. This can be done either:

(a) by allocating national projections of these magnitudes among regions using the best allocators available, or

(b) by trend analysis or other "bottom-up" approaches.

Step 2: *Derivation of Initial Output Projections*

Given the final demand vector derived in Step 1, the interregional input-output model can be run in the standard manner to yield a projection of output levels by sector by region in the year $t+\theta$. That is:

$$X_{t+\theta} = [I-\mathcal{A}E]^{-1} Y_{t+\theta}$$

Step 3: *Derivation of Initial Employment Demand Projections*

Given the output level projections derived in Step 2, labour requirements (employment demand) by sector by region can be projected via the use of a set of "reasonable" output to employment coefficients ($^J w_i$ and $^J w_{id}$) as in equations (6.2.17) and (6.2.18).

Step 4: *Derivation of Initial Population Projections*

Given the regional employment demand projections derived in Step 3, population by region can be projected via the use of a set of "reasonable" employment demand to population coefficients ($^J\varrho$ or $^J\varrho_d$; $d=1,\ldots,d$, $J=A,\ldots,U$). That is,

$$^J\underline{P}_{t+\theta} = {}^J\varrho_D \, {}^J\underline{E}_{t+\theta}$$

or

$$^J P_{t+\theta} = \sum_d {}^J Q_D \, {}^J E_{d,t+\theta}$$

Step 5: *Derivation of Regional Wage Income Projections*

The next step involves the derivation of crude projections of regional wage and salary incomes for the year $t+\theta$. This is done for each region by (a) multiplying each sector's output level projection derived at the end of Step 2 by the dollar's worth of labour required per dollar output (i.e. by the input-output coefficient $^{JJ}a_{Lj}$ in that sector's wage and salary row), and then (b) summing over all sectors.

Step 6: *Derivation of Household Disposable Income Projections*

The next step involves the use of econometric equations (6.2.3) - (6.2.12) to derive projections for:

(a) non-wage income by type by region $^J NWY_k$ (k=1,...,6);

(b) household income $^J HY$ by region -- as an identity from (a) and the results of Step 5;

(c) household disposable income $^J HDY$ by region;

(d) household disposable income per capita $^J HDY$ by region -- as an identity from (c) and the results of Step 4.

Step 7: *Derivation of Regional Consumption Expenditure Projections by Type*

Use of the results of Step 4 (population levels) and Step 6 (household disposable income) in the econometric equations (6.2.13) - (6.2.16) to derive projections for household consumption expenditure by type by region.

Step 8: *Derviation of Regional Consumption Expenditure Projections by Sector*

The next step involves the use of a transformation matrix, designated T_1, to convert the consumption expenditure projections derived in Step 7 into the format required for the input-output module -- that is, into household consumption expenditure by sector by region.[34]

Step 9: *Revision of Regional Output Projections*

The household consumption expenditure projections derived at the end of Step 8 are used to revise the final demand vector employed when running the input-output module in Step 2.

Steps 2-9 are then repeated for as many rounds as is required to reach a reasonable level of convergence.

6.3.2 Linkage with Government Expenditure Submodel

Steps 1-8:

As in Steps 1-8 of Section 6.3.1.

Step 9: *Derivation of Regional Government Expenditure Projections by Type*

The next step involves the use of the results of Step 4 (population), Step 6 (household disposable income) and Step 2 (output by sector) in the econometric equations (6.2.33) - (6.2.45) to derive projections for government expenditure by type by level by region.

Step 10: *Derivation of Regional Government Expenditure Projections by Sector*

The next step involves the use of a transformation matrix, designated T_2, to convert the government expenditure projections derived in Step 9 into the format required for the input-output module -- that is, government expenditure by sector by region.[35]

Step 11: *Revision of Regional Output Projections*

The household consumption expenditure projections derived at the end of Step 8, and the government expenditure projections derived at the end of Step 10 are then used to revise the final demand vector employed when running the input-output module in Step 2.

Steps 2-11 are then repeated for as many rounds as is required to reach a reasonable level of convergence.

6.3.3 Linkage with Labour Market Submodel

Step 1-4:

As in Steps 1-4 of Section 6.3.1. This yields output, employment demand and population projections for each region.

Step 5: *Derivation of Initial Labour Supply Projections*

The next step involves the use of the set of econometric equations (6.2.20) - (6.2.24) to derive regional labour supply projections for year $t+\theta$.

Step 6: *Derivation of Regional Wage Rate Projections*

The next step involves the use of (a) the employment demand projections derived in Step 3, and (b) the labour supply projections derived in Step 5, in the econometric equations (6.2.25) - (6.2.29) to get regional wage rate projections for year $t+\theta$.

Step 7: *Derivation of Regional Wage Income Projections*

Regional wage income projections can then be made via the use of the identity in equation (6.2.1) -- that is, by multiplying the employment demand projections derived in Step 3 by the wage rate projections derived in Step 6, and then summing over all sectors (and demographic classes).

Steps 8-10:

As in Steps 6-8 of Section 6.3.1. This yields household disposable income and household consumption expenditure projections for each region.

Steps 11-12:

As in Steps 9-10 of Section 6.3.2. This yields government expenditure projections for each region.

Step 13: *Revision of Regional Output Projections*

The household consumption expenditure projections derived at the end of Step 10, and the government expenditure projections derived at the end of Step 12 are then used to revise the final demand vector employed in the input-output module in Step 2.

Steps 2-13 are then repeated for as many rounds as is required to achieve a reasonable level of convergence.

6.3.4 Linkage with Private Investment Expenditure Submodel

Steps 1-4:

As in Steps 1-4 of Section 6.3.1. This yields output, employment demand and population projections for each region.

Steps 5-7:

As in Steps 5-7 of Section 6.3.3. This yields wage income projections for each region.

Steps 8-10:

As in Steps 6-8 of Section 6.3.1. This yields household disposable income and household consumption expenditure projections for each region.

Steps 11-12:

As in Steps 9-10 of Section 6.3.2. This yields government expenditure projections for each region.

Step 13: *Derivation of Regional Private Investment Expenditure Projections by Purchasing Sector*

The next step involves the use of the results of Step 2 (output by sector), Step 4 (population), Step 6 (wage rates), Step 7 (wage income) and Step 8 (household disposable income) in the econometric equations (6.2.30) - (6.2.32) to derive projections for private investment expenditure by sector making the purchase and region in which the investment is to be carried out.

Step 14: *Derivation of Regional Private Investment Expenditure Projections by Supplying Sector*

The next step involves the use of a transformation matrix, designated T_3, to convert the private investment expenditure projections derived in Step 13 into the format required for the input-output module -- that is into private investment expenditure by sector and region which supplies the goods and services required to carry out the investment.[36]

Step 15: *Revision of Regional Output Projections*

The household consumption expenditure projections derived at the end of Step 10, the government expenditure projections derived at the end of Step 12, and the private investment expenditure projections derived at the end of Step 14 are then used to revise the final demand vector employed in the input-output module in Step 2.

Steps 2-15 are then repeated for as many rounds as required to achieve a reasonable level of convergence.

6.4 Integration of Multiregional Input-Output, Demographic and Econometric Modules[37]

The linkage of multregional input-output and demographic modules was discussed in Section 5.4. However, here not only were the final demand vectors required for operation of the input-output module taken as given, but in addition the regional wage rates (income levels) required for operation of the demographic module were taken as fixed. This suggests that the addition of a multiregional econometric module of the form discussed in Section 6.2 will produce superior economic and demographic projections. The basic steps involved in operating this more complete model are as follows:

Steps 1-5:

As in Steps 1-5 of Section 5.4, this involves deriving initial population, final demand and output projections.

Step 6: *Derivation of Initial Employment Demand Projections*

Using the results of Step 5 (output level projections), as in Step 3 of Section 6.3.1.

Step 7: *Derivation of Initial Labour Supply Projections*

Using the results of Step 2 (population level projections), as in Step 5 of Section 6.3.3.

Step 8-9: *Derivation of Regional Wage Rate and Wage Income Projections*

Using the results of Steps 6 (regional employment demands) and Step 7 (regional labour supplies), as in Steps 6 and 7 of Section 6.3.3.

Steps 10-12: *Derivation of Gross Outmigration and Inmigration Rate Projections*

Using the results of Step 6 (regional employment demands), Step 7 (regional labor supplies) and Step 9 (regional wage and salary incomes), as in Steps 9-11 of Section 5.4.

Steps 13-15: *Derivation of Net Interregional Migration Flow Projections*

Using the results of Step 12 (gross regional migration levels), as in Steps 12-14 of Section 5.4. These estimates of the economically induced change in each region's population over the projection period are then used to revise the initial crude population projections derived for each region at the end of Step 2.

Step 16: *Derivation of Household Disposable Income Projections*

Using the results of Steps 9 (regional wage and salary income) and 15 (revised regional population levels), as in Step 6 of Section 6.3.1.

Step 17: *Derivation of Regional Consumption Expenditure Projections*

Using the results of Steps 15 (revised regional population levels) and 16 (household disposable income), as in Steps 7 and 8 of Section 6.3.1.

Step 18: *Derivation of Regional Government Expenditure Projections*

Using the results of Steps 15 (revised population levels), 16 (household disposable income) and 5 (output by sector), as in Steps 9 and 10 of Section 6.3.2.

Step 19: *Derivation of Regional Private Investment Expenditure Projections*

Using the results of Steps 5 (output by sector), 15 (revised population levels), 8 (regional wage rates) and 9 (regional wage and salary income), as in Steps 13 and 14 of Section 6.3.4.

Step 20: *Revision of Regional Output Projections*

The household consumption expenditure projections derived at the end of Step 17, the government expenditure projections derived at the end of Step 18, and the private investment expenditure projections derived at the end of Step 19 are then used to revise the final demand vector employed in the input-output module in Step 5.

Steps 5-20 are then repeated for as many rounds as is required to achieve a reasonable level of convergence.

6.5 Integration of Multiregional Input-Output, Comparative Cost-Industrial Complex, Programming, Demographic and Econometric Modules

A schematic representation of the nature of this more complete integration is given in Figure 6.5. The basic steps involved are:

Step 1: *Derivation of Initial Population Projections*

As in Steps 1-2 of Section 5.4

Step 2: *Identification of Sectoral Structure and Setting Up of Modified Interregional Input-Output Model*

As in Step 3 of Section 5.5.

Step 3: *Derivation of Initial Traditional Final Demand*

Using the results of Step 2 (regional population levels), as in Step 4 of Section 5.5.

Step 4: *Derivation of Initial Non-Traditional Final Demand Projections*

As in Steps 5 and 6 of Section 5.5.

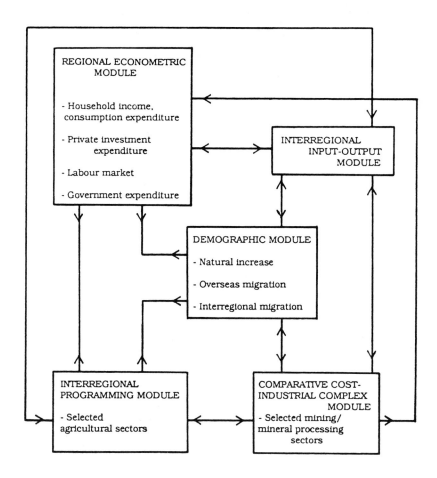

Figure 6.5 Linkage of Input-Output, Comparative Cost-Industrial Complex, Programming, Demographic and Econometric Modules

Step 5: *Consolidation of Traditional and Non-Traditional Final Demand Projections*

Using the results of Steps 3 (traditional final demand) and 4 (non-traditional final demand), as in Step 7 of Section 5.5.

Step 6: *Derivation of Consistent Output Projections for Energy Product,Energy Supply and Other Non-Energy Sectors*

A consistent set of output projections can be obtained for energy product P, energy supply S and non-energy N sectors via use of the results of Steps 5 (final demand projections) and 2 (modified input-output coefficients matrix) as data inputs into Steps 8-11 of Section 5.5.

Step 7: *Elimination of Inconsistencies between Cost Sensitive and Other Output Projections*

The output vectors X_N', X_N'', X_P and X_S derived from a first run-through of Steps 1-6 can be expected to contain a number of inconsistencies of the type described in Step 12 of Section 4.4.2

If any of these inconsistencies are judged to be too large to be permitted in the final projections, the analyst must re-estimate one or more of:

(a) the level of traditional final demand by sector by region;

(b) the regional pattern of industrial markets (intermediate demands) for the output of cost-sensitive non-energy sectors; and

(c) the regional cost differentials (especially re: energy) associated with the operation of one or more cost-sensitive non-energy sectors.

The results of Steps 5 and 6 provide the basic materials required to perform the re-estimations in (b) and (c) above; while those in (a) are performed via the use of Steps 8-17 below.

ECONOMETRIC MODULE 243

Step 8: *Derivation of Initial Employment Demand Projections*

Using the results of Step 7 (output level projections), as in Step 3 of Section 6.3.1.

Step 9: *Derivation of Initial Labour Supply Projections*

Using the results of Step 2 (population level projections), as in Step 5 of Section 6.3.3.

Step 10: *Derivation of Regional Wage Rate and Wage Income Projections*

Using the results of Steps 8 (regional employment demands) and 9 (regional labour supplies), as in Steps 6 and 7 of Section 6.3.3.

Step 11: *Derivation of Gross Outmigration and Inmigration Projections*

Using the results of Steps 8 (regional employment demands), 9 (regional labour supplies) and 10 (regional wage and salary incomes), as in Steps 9-11 of Section 5.4.

Step 12: *Derivation of Net Interregional Migration Flow Projections*

Using the results of Step 11 (gross regional migration levels), as in Steps 12-14 of Section 5.4. These estimates of the economically induced change in each region's population over the projection period can then be employed to revise the initial crude population projections derived at the end of Step 2.

Step 13: *Derivation of Regional Household Disposable Income Projections*

Using the results of Steps 10 (regional wage and salary income) and 12 (revised regional population levels), as in Step 6 of Section 6.3.1.

Step 14: *Derivation of Regional Consumption Expenditure Projections*

Using the results of Steps 12 (revised regional population levels) and 13 (household disposable income), as in Steps 7 and 8 of Section 6.3.1.

Step 15: *Derivation of Regional Government Expenditure Projections*

Using the results of Step 7 (output by sector), 12 (revised population levels) and 13 (household disposable incomes) as in Steps 9 and 10 of Section 6.3.2.

Step 16: *Derivation of Regional Private Investment Expenditure Projections*

Using the results of Steps 7 (output by sector), 12 (revised population levels) and 10 (regional wage rates and incomes), as in Steps 13 and 14 of Section 6.3.4.

Step 17: *Rerun of Integrated Model*

The household consumption expenditure projections derived at the end of Step 14, the government expenditure projections derived at the end of Step 15, and the private investment expenditure projections derived at the end of Step 16 are used to revise the initial traditional final demand projections made in Step 3. Steps 5-17 are then repeated, incorporating as well:

(a) the revised industrial markets for the output of cost-sensitive non-energy sectors derived at the end of Step 7;

(b) the revised cost differentials (especially re: energy) derived at the end of Step 7;

(c) the revised wage rates derived at the end of Step 10, when conducting new comparative cost and industrial complex analyses in Steps 5 and 6; and

(d) the revised population projections derived at the end of Step 12, when reestimating labour supply projections in Step 9 and persons "at risk" of migrating in Step 11.

A number of such reruns may be required until all major discrepancies are eliminated -- that is, until the integrated model converges on a reasonable set of results.

6.6 Concluding Remarks

This chapter has dealt with the econometric module and its linkage to the input-output, comparative cost-industrial complex, linear programming, and demographic modules of an integrated multiregion model.

The equations comprising this module have been specified in such a way that data is available to support their implementation in the Australian context. However when confronted with this data, these equations could prove virtually useless for projection purposes. As a result we need to consider the use of one or more fallback assumptions which would enable the integrated model to be operated despite the elimination of such equations. For example, we may:

(1) assume fixed wage rates in each region, thereby eliminating the need for equations (6.2.25) - (6.2.28);

(2) assume fixed labour force participation rates in each region, thereby eliminating the need for equations (6.2.21) - (6.2.23);

(3) assume a fixed relationship between wage income and total household income for each region, thereby eliminating the need for equations (6.2.3) - (6.2.9);

(4) assume a fixed relationship between consumption expenditures and household disposable income levels for each region, thereby eliminating the need for equations (6.2.14) - (6.2.16);

(5) assume a fixed relationship between output levels and private investment expenditures in each region, thereby eliminating the need for equations (6.2.30) - (6.2.32); and

(6) assume that state and local government expenditures in each region increase (or decrease) in proportion to a set of reasonable indicators -- such indicators (eg. population, output levels, income levels) being endogenous to other components of the integrated model. This then eliminates the need for equations (6.2.34) - (6.2.45).

NOTES

[1] In this figure, bold boxes indicate variables endogenous to the household income and consumption expenditure submodel; dashed boxes indicate variables endogenous to the multiregional econometric module, but exogenous to this particular submodel; ovals indicate variables endogenous to the overall integrated model, but exogenous to the multiregional econometric module; and circles indicate variables exogenous to the overall integrated model.
Note that it is recognised that some of the specifications given below when describing the structure of each submodel might be considerably changed and improved after being confronted with empirical material.

[2] Other multiregional econometric models which contain equations for determining household income include those by Crow (1973), Ballard and Glickman (1977), Harris and Nadji (1980), Milne, Glickman and Adams (1980) and Ballard and Wendling (1980).

[3] Due to data limitations in the Australian context, we have i = 1,...,4 (where 1=defence, 2=agriculture, 3=manufacturing, 4=other non-manufacturing), J=1,...,6 (where 1=NSW + ACT, 2=VICT, 3=QLD, 4=SA + NT, 5=WA, 6=TAS), and d=1,...,4 (where 1=male aged 15-19, 2=female aged 15-19, 3=male aged 20+, and 4=female aged 20+).

[4] The proportions $^J\delta_i$ (i=1,...,n) can be calculated from base year data and assumed to remain unchanged over the projection period. Alternatively, the analyst may wish to adjust one or more of these proportions to reflect changes that are expected to occur over this period.

[5] The latter variable is included to reflect a partial (ie. lagged) adjustment of dwelling rents to changes in effective demand for housing.

[6] This category comprises (1) interest on life and superannuation funds (imputed), (2) other interest and dividends, (3) third party insurance transfers, and (4) transfers from overseas.

[7] More rigorously, we deduct income used to make the following payments (1) income taxes, (2) other direct taxes, fees, fines, etc., (3) consumer debt interest, and (4) net transfers overseas.

[8] The proportions $^J\psi$ (J=A,...,U) can be calculated from base year data and assumed to remain constant over the projection period. Alternatively, the analyst may wish to adjust one or more of these proportions to reflect changes that are expected to occur over this period.

[9] Other multiregional econometric models which contain equations for determining household consumption expenditures include those by Brown, et al (1972), Crow (1973), and Harris and Nadji (1980).

[10] That is, although consumer expenditure can be expected to respond to changes in income levels, unemployment rates and consumer credit conditions, it often does so with a lag.

[11] The latter variable is included to reflect a partial (lagged) adjustment of dwelling rents to changes in effective demand for housing. Note the similarity between the form of this equation and that for income from dwellings (namely 6.2.7). This is so since in general we would expect these two dependent variables to differ only by the extent to which: (1) non-residents operate rental housing in the given region, and (2) residents of the given region operate rental housing in other regions.

[12] Although these coefficients must be fixed in the sense that they do not change with the level of output in sector i, J, they can be adjusted over time should reasonable indicators of the nature of the required changes be available. See Chapter 7 for relevant discussion.

[13] For a model employing a similar set of equations, see Fromm, et al (1979). Other multiregional econometric models containing employment demand equations include Milne, et al (1980), Crow (1973), Ballard, et al (1980a,1980b), de Falleur, et al (1981), and Glejser, et al (1973).

[14] Note that the user cost of capital is assumed not to vary between regions -- in part because capital markets operate nationally rather than regionally in the Australian context resulting in minimal (if any) regional variations in interest rates, but also because of data limitations.

[15] Due to data limitations in the Australian context, we have d=1,...,12 (where 1=males aged 15-19, 2=males aged 20-24, 3=males aged 25-34, 4=males aged 35-44, 5=males aged 45-64, 6=males aged 65+, 7=females aged 15-19, 8=females aged 20-24, and so on).

The values for $^{J}H_{dt}$ (d=1,...,d) can be calculated from base year data and assumed to remain unchanged over the projection period. Alternatively, the analyst may wish to adjust one or more of these values to reflect changes that are expected to occur over this period.

[16] Other multiregional econometric modules which contain equations for determining labour force participation rates are Courbis(1975, 1979, 1980), de Falleur, et al (1981) and Fromm, et al (1979,1980).
Note that when estimating equations (6.2.21) - (6.2.23), experiments will be conducted with multinomial logit versions as well as linear versions

of these equations. However, because of data limitations which require pooling of time series and cross section data the latter format may be expected to perform better and provide more easily interpreted results.

[17] Note that the unemployment level (measured in terms of number of persons) $^J UN_{dt}$ can be calculated as $(_S {}^J E_{dt} - {}_D {}^J E_{dt})/{}^J H_{dt}$.

[18] The particular index chosen is the proportion of total labour requirements in the given region that is met using female employees. Hence

$$_D {}^J E_{Ft} = \sum_{d=1}^{12} {}_D {}^J E_{Fdt}$$

[19] Other multiregional econometric models which contain equations for determining regional wage rates include Ballard and Glickman (1977), Fromm, et al (1979, 1980), Milne, et al (1980, 1980a), Courbis (1975, 1979, 1980), Crow (1973) and Ballard and Wendling (1980).

[20] Once again due to data limitations in the Australian context we also have J=1,...,6 and d=1,...,4. See footnote 3 above, for further details on the nature of regions and demographic classes employed.

Much empirical research in the Australian context has revealed strong evidence to suggest a distinct segmentation of the labour market via sex and age (wherein the age categories are 'youth' versus 'non-youth').

[21] The particular productivity measure chosen is the value added VA_i per manhour and employment $_D E_i$ in sector i.

[22] In general, we would expect the sign of h_1 to be positive -- that is the 'tighter' the labour market for i,d in region J, the smaller the basic wage component of the corresponding total wage rate.

[23] For example, where the 'excess' supply of labour available to agricultural producers is large relative to their manpower requirements, then wage rate increases in the agricultural sector may lag behind those in other sectors. On the other hand, should available supply be small, then wage rates offered by agricultural producers will need to be competitive with those in other sectors if their manpower requirements are to be met.

[24] That is for each region J (J=A,...,U) we have

$$\frac{^JWR_{rt} - {}^JWR_{rt-1}}{^JWR_{rt-1}} = \frac{^JWR_{st} - {}^JWR_{st-1}}{^JWR_{st-1}} \qquad (6.2.29)$$

where r = defence sectors and s = civilian sectors.

[25] Other multiregional econometric models which contain equations for determining gross fixed capital expenditures include Brown, et al (1972), Crow (1973, 1979), Harris (1972, 1973, 1980a, 1980b), de Falleur, et al (1975) and Lakshmanan (1980, 1982).

[26] Note that this variable is included as an indirect way of identifying regional differences in the user cost of capital due to government policy -- since the proxy variable chosen to represent the UCC, namely the interest rate on government bonds, is only available for the nation as a whole.

[27] Note that the demand for housing is related not only to per capita income levels and population growth but also to household formation processes. The latter variable is not included in equation (6.2.31) due to data limitations at the regional level.

[28] Other multiregional econometric models which contain equations for determining the level of federal government expenditures by region include Crow (1973), Brown, et al (1972), and Harris and Nadji (1980).

[29] In the Australian context we let g=1 federal, 2 state and 3 local, while p=1 public administration, 2 education, 3 health, 4 social security and welfare, 5 housing and community amenities, 6 recreation and culture, 7 economic services, and 8 defence.
Since all defence expenditures are treated as current in the Australian national accounts the functions q (q=1,...,7) are identical in character to those described above.

[30] The constants $^J\psi_{pt}$ (p=1,...,8) and $^J\Omega_{qt}$ (q=1,...,7) can be derived from base year data and assumed to remain unchanged over the projection period. (See the discussion in Smith (1983, Section 3.3). Alternatively, the analyst may wish to adjust these allocators for one or more functions to reflect changes that are expected (or projected) to occur over this period. See the discussion on the use of an interdependent policy formation module in Chapter 9.

[31] Other multiregional econometric models which contain equations for determining the level of state and/or local government expenditures by

region include Crow (1973), Ballard and Glickman (1977), and Ballard and Wendling (1980).

[32] In the Australian context we let i = 1 agriculture, forestry and fishing services, 2=mining, manufacturing and construction services, 3 = electricity, gas and water services, 4 = transportation services and 5 = other.

[33] Other multregional models which involve the integration of input-output and econometric modules include Courbis (1975, 1979, 1980), Funck and Rembold (1975), Treyz and Stevens (1980) and Isard and Anselin (1982).

[34] If n = the number of input-output sectors, U = number of regions, and q = number of consumption expenditure types, then T_1 is a (qxUn) matrix. A typical element of this matrix shows the proportion of household consumption expenditure of type q (q=1,...,q) which is purchased from sector i (i=1,...,n) in region J (J=A,...,U). One T_1 matrix is required for each region, and can be derived from base year data and the corresponding national transformation matrix developed by the input-output section of the Australian Bureau of Statistics using the procedures described in Smith (1983, Section 3.3.1).

[35] If n = the number of input-output sectors, U = the number of regions, p = number of government expenditure types and g = number of levels of government, then T_2 is a (pgxUn) matrix. A typical element of this matrix shows the proportion of expenditure of type p (p=1,...,p) made by level of government g (g=1,...,g) which is spent on purchases from sector i (i=1,...,n) in region J (J=A,...,U). One T_2 matrix is required for each region, and can be derived from base-year data and the corresponding national transformation matrix developed by the input-output section of the Australian Bureau of Statistics using the procedures described in Smith (1983), Sections 3.3.2 and 3.3.4).

[36] If n = the number of input-output sectors, and U = number of regions, the T_3 is a (nxUn) matrix. A typical element of this matrix shows the proportion of sector i's investment expenditure (i=1,...,n) which is spent on purchases from sector j (j=1,...,n) in region J (J=A,...,U). One T_3 matrix is required for each region, and can be derived from base-year data and the corresponding national transformation matrix developed by the input-output section of the Australian Bureau of Statistics using the procedures described in Smith (1983, Section 3.3.3).

[37] Other multiregional models which involve the integration of input-output, demographic and econometric modules include Courbis (1975,

1979, 1980), Funck and Rembold (1975), Treyz and Stevens (1980) and Isard and Anselin (1982).

CHAPTER 7
A FACTOR DEMAND MODULE AS A COMPONENT OF AN INTEGRATED MULTIREGIONAL MODEL

7.1 Introduction

This chapter is concerned with the development of a factor demand (substitution) module as a component of an integrated multiregional model. In Section 7.2 we discuss two alternative functional forms -- the translog and Diewert (generalized Leontief) -- which can be used to project substitution between factor inputs (capital, labour, energy and materials) given the corresponding factor price changes. In Section 7.3, we explore the relationship between the cost shares and/or factor demand levels derived from these factor demand models and the technical-trade coefficients comprising the interregional input-output module. Given this relationship, Section 7.4 develops sets of procedures which may be used to revise the input-output technical-trade coefficients given the factor substitution patterns projected using the factor demand module.

In Section 7.5 we outline the steps involved in integrating multiregional input-output, econometric and factor demand modules; and in Section 7.6 the steps involved in integrating multiregional input-output, demographic and factor demand modules. The operation of a more fully integrated model comprising: (a) an interregional input-output module, (b) a comparative cost-industrial complex module, (c) an interregional programming module, (d) a multiregional demographic module, (e) a regional econometric module, and (f) a factor demand module is discussed in Section 7.7. Finally, some concluding remarks are made in Section 7.8.

7.2 Outline of Factor Demand Models

This section outlines the potential for development of a factor demand module capable of incorporation as a component of an integrated multiregional model.[1] Until recently, the development of factor demand models was limited since available production function forms (e.g. Leontief, Cobb Douglas, and CES) implied quite restrictive prior assumptions pertaining to technical production relations.[2] However, some recent developments in microeconomic theory have led to important advances in the state-of-the-art re: factor demand modelling.

In particular, the development of more general and complex functional forms -- for example, the Diewert (generalized Leontief), the generalized CES, the translog (transcedental logarithmic) and the generalized Box-Cox[3] -- has expanded greatly the scope for development of factor demand models capable of exploring substitution effects caused by relative price changes. The simultaneous development of duality theory[4] which given certain properties of the underlying production structure[5] allows the analyst to derive equilibrium factor demand levels from cost functions directly[6] -- has permitted greater use of these more flexible forms in operational models.[7]

In what follows, we have restricted our attention to those factor demand models which employ either Diewert or translog cost functions. This is so because our primary objective is the establishment of reasonable sets of linkages between factor demand, input-output and other multiregional models. It is hoped that, given the principles employed in developing these linkages, the reader can then make the necessary modifications should he choose to employ one of the other functional forms mentioned here.

7.2.1 Translog Cost Function

The point of departure for our factor demand module is a twice differentiable, well behaved (regular) production structure which is weakly separable in four factors: capital (K), labour (L), energy (E) and materials (M):[8]

$$Q_i = F(Q_{Ki}, Q_{Ei}, Q_{Li}, Q_{Mi}) \qquad (7.2.1)$$

where Q_i is the output level in sector i, and Q_{ri} (r= K,E,L,M) are the factor input levels for sector i. In order to keep things simple at this stage, we assume that the production function in (7.2.1) exhibits constant returns to scale.[9] Then by the use of the duality theorem and a translog functional form, the corresponding unit cost function is given by:[10]

$$\ln C_i = \alpha_{0i} + \sum_r \alpha_{ri} \ln P_r + 1/2 \sum_{r,s} \beta_{rs,i} \ln P_r \ln P_s$$
$$(7.2.2)$$

where r, s = K,L,E,M; C = production cost per unit output in sector i; and P_r = price (cost) per unit of input r.

A convenient feature of the cost function approach is that the derived demand functions for the factors of production can be easily computed by partially differentiating the cost function with respect to factor prices. That is, from Shepard's (1953) lemma:

$$\delta TC_i / \delta P_i = Q_{ri}$$

or $\qquad (7.2.3)$

$$(\delta C_i / \delta P_r) = q_{ri}[11]$$

where TC_i = total production costs in sector i; Q_{ri} = the cost minimizing quantity of factor input r for sector i; and q_{ri} = the cost minimizing quantity of r per unit output of sector i.

Further, since the cost function is linear homogeneous in prices $C_i = \Sigma P_r q_{ri}$ and equation (7.2.3) can be expressed in logarithmic form for the translog unit cost function as:

$$\frac{\delta \ln C_i}{\delta \ln P_i} = \frac{\delta C_i}{\delta P_i} \cdot \frac{P_r}{C_r} = \frac{P_r q_{ri}}{\sum_r P_r q_{ri}} = S_{ri}$$

then S_{ri} is the share of factor r in total production costs, and is given by:

$$S_{ri} = \alpha_{ri} + \sum_s \beta_{rs,i} \ln P_s \quad (r,s=K,L,E,M) \tag{7.2.4}$$

The criterion of shares adding up to unity and the properties of neoclassical production theory require that the following restrictions be placed on (7.2.4) for r,s = K.L.E.M:[12]

$$\sum_r \alpha_{ri} = 1; \quad \sum_r \beta_{rs,i} = \sum_s \beta_{rs,i} = 0; \text{ and } \beta_{rs,i} = \beta_{sr,i} \quad r \neq s \tag{7.2.5}$$

The stochastic specification of the model in (7.2.4)-(7.2.5) is completed by appending additive error terms μ_{ri} to the share equations in (7.2.4).[13] That is:

$$S_{ri} = \alpha_{ri} + \sum_s \beta_{rs,i} \ln P_s + \mu_{ri} \tag{7.2.6}$$

With coefficients a_{ri} and $\beta_{rs,i}$ estimated on the basis of past data, it then becomes possible to project the cost shares S_{ri} (r = K,L,E,M) for any given set of factor input prices P_K, P_L, P_E and P_M.[14] Further, as will be shown in Section 7.3.1, for any given sector i and region J, the cost share $^JS_{ri}$ is equivalent to the input-output coefficient $^Ja_{ri}$ expressed in dollar units.

7.2.2 Diewert (Generalized Leontief) Cost Function

When a Diewert functional form is employed, the unit cost function corresponding to equation (7.2.2) is given by:[15]

$$C_i = \sum_r \sum_s \gamma_{rs,i} (P_r/P_s)^{1/2} \qquad (7.2.7)$$

where r, s = K,L,E,M and $g_{rs} = g_{sr}$, $r \neq s$.

Thus use of Shepard's (1953) lemma now gives:[16]

$$(\delta C_i/\delta P_r) = q_{ri} = \sum_s \gamma_{rs,i} (P_r/P_s)^{1/2} \qquad (7.2.8)$$

or

$$Q_{ri} = \sum_s \gamma_{rs,i} (P_r/P_s)^{1/2} Q_i \qquad (7.2.9)$$

where as before q_{ri} = the cost minimising quantity of factor input r per unit output; Q_{ri} = the cost minimising quantity of r; and Q_i = level of output in the given sector.

The stochastic representation of the model in (7.2.8) is completed by appending an additive error term μ_{ri}. That is

$$q_{ri} = \sum_s \gamma_{rs,i} (P_r/P_s)^{1/2} + \mu_{ri} \qquad (7.2.10)$$

With coefficients $\gamma_{rs,i}$ estimated on the basis of past data, it then becomes possible to project the factor demands per unit output q_{ri} (r=K,L,E,M) for any given set of factor prices P_K, P_L, P_E, and P_M. Further, as will be shown below, for any given sector i and region J the derived demand $^J q_{ri}$ can be converted into the input-output coefficient

$^J a_{ri}$ by use of the price ratio ($^J P_i / ^J P_r$) where $^J P_i$ is the price per unit output in i, J.

7.2.3 Further Refinements

The form of the factor demand module outlined in sections 7.2.1 and 7.2.2 will be that referred to below when developing a set of reasonable linkages to other modules within an integrated multiregional model. However, it is useful to point up two further refinements which may be explored once the simpler set of linkages has been made operational.

a. Recognition of Economies of Scale[17]

The translog cost function can be used to describe production structures which exhibit non-constant returns to scale. When this is the case, the unit cost function in equation (7.2.2) is replaced by a total cost function of the form:

$$\ln TC = \alpha_o + \alpha_Q \ln Q + 1/2\, \beta_Q (\ln Q)^2 + \sum_r \alpha_r \ln P_r$$

(7.2.11)

$$+ 1/2 \sum_r \sum_s \beta_{rs} \ln P_r \ln P_s + \sum_r \beta_{Qr} \ln Q \ln P$$

The cost share equations in (7.2.4) are then replaced by those of the form:

$$S_r = \alpha_r + \sum_s \beta_{rs} \ln P_s + \beta_{Qr} \ln Q$$

(7.2.12)

In addition to the restrictions given in (7.2.5), it must now be that

$$\sum_r \beta_{Qr} = 1 \qquad r=K,L,E,M$$

(7.2.13)

Equations (7.2.12), (7.2.13) and (7.2.5) can once again be transformed into a stochastic model by the appendage of an additive error term to equation (7.2.12). That is

$$S_r = \alpha_r + \sum_s \beta_{rs} \ln P_s + \beta_{Qr} \ln Q + \mu_r \qquad (7.2.14)$$

The parameters α_r, β_{Qr} and β_{rs} (r,s=K,L,E,M) can then be estimated on the basis of past data using linear regression techniques. However, one now needs (exogenous) projections of sector output levels (Q_i, i=1,...,n) as well as factor demand prices (P_r, r=K,L,E,M) in order to derive forecasts of the corresponding cost shares (S_{ri}, r=K,L,E,M). But, since the cost shares S_{ri} represent input-output coefficients expressed in dollar terms and consistent sectoral output level projections cannot generally be made without knowledge of these coefficients, this suggests that the factor demand model given by (7.2.13), (7.2.14) and (7.2.5) should be operated iteratively with an input-output model in order to yield internally consistent projections of factors inputs and output levels. Since the iterative process (when carried out in a conceptually satisfying manner) would complicate the operation of the integrated model considerably, it is not clear that the additional refinement would be worth the costs involved. This is especially true given the quality of the data available for implementation of the factor demand model, and the level of accuracy which could justifiably be claimed for the base year input-output coefficients.

b. Recognition of Partial Adjustment Process[18]

The factor demand models discussed above are based on a comparative static view of factor input responses to relative price changes. That is, instantaneous adjustment is assumed.[19] Relaxing this assumption and allowing for disequilibrium processes suggests the introduction of a partial adjustment model. For example, when the

Diewert cost function is employed we may set the q_r derived from the use of equation (7.2.10) for projection purposes as the desired or equilibrium demand $*q_r(t+\theta)$ for factor r at time $t+\theta$. That is:[20]

$$*q_r(t+\theta) = q_r(t+\theta) = \sum_s \gamma_{rs,i} \, (P_r(t+\theta)/P_s(t+\theta))^{1/2} \tag{7.2.15}$$

The actual demand $q_r(t+\theta)$ for factor r at time $t+\theta$ can then be derived via the use of the relation:

$$q_r(t+\theta) - q_r(t) = \lambda_r \, [*q_r(t+\theta) - q_r(t)] \tag{7.2.16}$$

where q_r = the actual demand for factor r at time t; and

λ_r = a parameter defining the speed with which the actual demand for r adjusts to the desired levels ($0 < \lambda_r < 1$).

However, the critical problem then becomes how to determine the appropriate values for the parameters λ_r (r=K,L,E,M) -- particularly when past observations refer to actual rather than desired q_r coefficients. This is a problem which has not yet been resolved (see Almon, et al 1979). As a result, subsequent sections will ignore the linkage problems that would be created by the introduction of such a refinement.

7.3 Relationship Between Factor Demand and Input-Output Models
7.3.1 Input-Output Model with Prices Explicit

Because (1) the version of the regional input-output model which has price relations explicit is not well known, and (2) some points relating to the linkage of input-output and factor demand models can be best understood without the complex interpretations that must be provided when the 'interregional' dimension is incorporated, the regional system of input-output equations will be discussed first.

a. Regional Input-Output Relations

The basic input-output relations for region J can be <u>expressed in monetary units</u> as:

(1) the value of output (supply) $^J X_i$ in sector i must equal the value of sales to intermediate and final demand sectors. That is:

$$^J X_i = \sum_{j=1}^{n} {}^J X_{ij} + {}^J Y_i \qquad i=1,\ldots,n \tag{7.3.1}$$

or

$$^J X_i = \sum_{j=1}^{n} {}^J a_{ij} \, {}^J X_j + {}^J Y_i \qquad i=1,\ldots,n \tag{7.3.2}$$

where $^J a_{ij}$ is an input-output coefficient defined as the value of sector i's output purchased by sector j in region J, per unit of output of the latter sector. That is $^J a_{ij} = {}^J X_{ij} / {}^J X_j$.

(2) the value of outlays $^J X_i$ in sector i (i=1,...,n) must equal the value of purchases from intermediate and primary input sectors. That is

$$^J X_i = \sum_{j=1}^{n} {}^J X_{ji} + \sum_{r=1}^{r} {}^J X_{ri} \tag{7.3.3}$$

or

$$^J X_i = \sum_{j=1}^{n} {}^J a_{ji} \, {}^J X_i + \sum_{r=1}^{r} {}^J a_{ri} \, {}^J X_i \tag{7.3.4}$$

where $^J a_{ri}$ is defined as the value of payments to factor r made by sector i in region J, per unit of output of that sector. That is $^J a_{ri} = {}^J X_{ri} / {}^J X_i$.

Making prices explicit, equations (7.3.1) and (7.3.3) can be rewritten, respectively, as

$$^JP_i\,^JQ_i = \sum_{j=1}^{n} {}^JP_i\,^JQ_{ij} + {}^JP_i\,^JQ_{iy} \tag{7.3.5}$$

and

$$^JP_i\,^JQ_i = \sum_{j=1}^{n} {}^JP_j\,^JQ_{ji} + \sum_{r=1}^{r} {}^JP_r\,^JQ_{ri} \tag{7.3.6}$$

where ^JP_i = price (unit cost) of the output produced by sector i in region J;

^JQ_i = physical quantity of output produced by sector i in region J;

$^JQ_{ij}$ = physical quantity of sector i's output that is purchased by sector j in region J;

$^JQ_{iy}$ = physical quantity of sector i's output that is purchased by final demand sectors in region J;

^JP_r = price (unit cost) of factor r included in primary input rows for region J; and

$^JQ_{ri}$ = physical quantity of factor r purchased by sector i in region J.

We now introduce (1) a set of input-output coefficients $^Jq_{ij}$ (i,j=1,...,n) defined in physical units as the quantity of sector i's output purchased by sector j in region J, per unit output of sector j. That is $^Jq_{ij} = {}^JQ_{ij}/{}^JQ_j$; and (2) a set of coefficients $^Jq_{ri}$ (r=1,...,r, i=1,...,n) defined as the physical quantity of factor r required as input to sector i in region j, per unit output of sector i. That is

$$^Jq_{ri} = {}^JQ_{ri}/{}^JQ_i \tag{7.3.7}$$

Equations (7.3.5) and (7.3.6) can then be rewritten as:

$$^{J}P_i {}^{J}Q_i = \sum_{j=1}^{n} {}^{J}P_i {}^{J}q_{ij} {}^{J}Q_j + {}^{J}P_i {}^{J}Q_{iy} \qquad (7.3.8)$$

and

$$^{J}P_i {}^{J}Q_i = \sum_{j=1}^{n} {}^{J}P_j {}^{J}q_{ji} {}^{J}Q_i + \sum_{r=1}^{r} {}^{J}P_r {}^{J}q_{ri} {}^{J}Q_i \qquad (7.3.9)$$

Dividing both sides of (7.3.8) by $^{J}P_i$, we now have the basic input-output relations for region J <u>expressed in physical units</u> as:

$$^{J}Q_i = \sum_{j=1}^{n} {}^{J}q_{ij} {}^{J}Q_j + {}^{J}Q_{iy} \qquad (7.3.10)$$

It is clear that the input-output coefficients $^{J}q_{ij}$ in (7.3.10) and $^{J}a_{ij}$ in (7.3.2) are related to one another. To see the exact nature of this relation, we start with the definitional equation

$$^{J}a_{ij} = {}^{J}X_{ij}/{}^{J}X_j \qquad (7.3.11)$$

Then, making use of the identities $^{J}X_{ij} = {}^{J}P_i {}^{J}Q_{ij}$ and $^{J}X_j = {}^{J}P_j {}^{J}Q_j$ we can rewrite (7.3.11) as

$$^{J}a_{ij} = {}^{J}P_i {}^{J}Q_{ij}/{}^{J}P_j {}^{J}Q_j = ({}^{J}P_i/{}^{J}P_j)({}^{J}Q_{ij}/{}^{J}Q_j) \qquad (7.3.12)$$

Since, by definition $^{J}q_{ij} = ({}^{J}Q_{ij}/{}^{J}Q_j)$ we can substitute this relation in (7.3.12) to derive

$$^{J}a_{ij} = ({}^{J}P_i/{}^{J}P_j) {}^{J}q_{ij} \qquad (7.3.13)$$

By a similar set of operations the $^J q_{ri}$ in (7.3.9) and $^J a_{ri}$ in (7.3.4) can be related to one another as

$$^J a_{ri} = (^J P_r / ^J P_i) \, ^J q_{ri} \tag{7.3.14}$$

Equations (7.3.13) and (7.3.14) are basic identities which shall prove useful in achieving a meaningful linkage between input-output and factor demand models.

b. Multiregional Input-Output Relations

As discussed on page 21 ideally the trade coefficients in the interregional input-output model should be allowed to vary depending on relative price changes in the different regions comprising the system. However, in order to keep the problem of integration as simple as possible at this stage we choose to ignore this aspect of reality. Instead, we take

(1) $^{JL}c_i$, the proportion of region L's requirements for sector i's output that is purchased from region J, to be constant for all $i=1,\ldots,n$ and $J,L=A,\ldots,U$; and

(2) $^{JL}c_r$, the proportion of region J's requirements for factor r that is purchased from region L, to be a constant for all $r=1,\ldots,r$ and $J,L=A,\ldots,U$.

When this is done, the basic input-output relations expressed in monetary units become

$$^J X_i = \sum_{j=1}^{n} \sum_{L=A}^{U} (^{JL}c_i)(^L a_{ij}) \, ^L X_j + \sum_{L=A}^{U} (^{JL}c_i) \, ^L Y_i \tag{7.3.15}$$

and

$$^J X_i = \sum_{i=1}^{n} \sum_{L=A}^{U} (^{LJ}c_j)(^J a_{ji}) \, ^J X_j + \sum_{r=1}^{r} \sum_{L=A}^{U} (^{LJ}c_r)(^J a_{ri}) \, ^J X_i \tag{7.3.16}$$

while the input-output relations experessed in physical units become

$$^JQ_i = \sum_{j=1}^{n} \sum_{L=A}^{U} (^{JL}c_i)(^Lq_{ij})^LQ_j + \sum_{L=A}^{U} (^{JL}c_i)^LQ_{iy}$$

and[21] (7.3.17)

$$^JP_i{}^JQ_i = \sum_{j=1}^{n} \sum_{L=A}^{U} {}^JP_i(^{LJ}c_i)(^Jq_{ji})^JQ_i + \sum_{r=1}^{r} \sum_{L=A}^{U} {}^JP_r(^{LJ}c_r)(^Jq_{ri})^JQ_i$$

7.3.2 Translog Cost Shares as Input-Output Coefficients

Let the endogenous producing sectors (i=1,...n) be partitioned into two sets, namely those producing material inputs (m=1,...,m) and those producing energy inputs (e=1,...,e). Further, let the primary input sectors be partitioned into two sets, namely those providing labour services (L) and those providing capital services (K). Then the relations in (7.3.3) and (7.3.4) can be rewritten, respectively, as:[22]

$$^JX_i = \sum_{m=1}^{m} {}^JX_{mi} + \sum_{e=1}^{e} {}^JX_{ei} + {}^JX_{Li} + {}^JX_{Ki}$$

(7.3.18)

and

$$^JX_i = \sum_{m=1}^{m} {}^Ja_{mi}{}^JX_i + \sum_{e=1}^{e} {}^Ja_{ei}{}^JX_i + {}^Ja_{Li}{}^JX_i + {}^Ja_{Ki}{}^JX_i$$

(7.3.19)

where $^Ja_{mi}$ = value of material input m purchased by sector i in region J per unit of its output;

$^Ja_{ei}$ = value of energy input e purchased by sector i in region J per unit of its output;

$^Ja_{Li}$ = value of labour services L purchased by sector i in region J per unit of its output; and

$^J a_{Ki}$ = value of capital services K purchased by sector i in region J per unit of its output.

Letting $^J a_{Mi} = \Sigma_m {}^J a_{mi}$ and $^J a_{Ei} = \Sigma_e {}^J a_{ei}$ equation (7.3.19) can be rewritten as

$$^J X_i = {}^J a_{Mi} {}^J X_i + {}^J a_{Ei} {}^J X_i + {}^J a_{Li} {}^J X_i + {}^J a_{Ki} {}^J X_i \qquad (7.3.20)$$

Since the RHS of (7.3.20) represents the total costs (outlays) by sector i in region J, it is clear that by definition the input-ouptut coefficients $^J a_{ri}$ (r=K,L,E,M) in (7.3.20) are equivalent to the cost shares $^J S_{ri}$ (r=K,L,E,M) derived from the use of a translog form for the unit cost function in the factor demand model. That is, from (7.3.11) and (7.3.18) we have

$$^J a_{ri} = {}^J X_{ri} / {}^J X_i = {}^J X_{ri} / \sum {}^J X_{ri} \qquad r=K,L,E,M \qquad (7.3.21)$$

or introducing prices

$$^J a_{ri} = {}^J P_r {}^J Q_{ri} / \sum {}^J P_r {}^J Q_{ri} \qquad r=K,L,E,M \qquad (7.3.22)$$

But from (7.2.4), the RHS of (7.3.22) is equivalent to the cost share $^J S_{ri}$.

7.3.3 Diewert Derived Factor Demands as Input-Output Coefficients

Employing the same partitioning of sectors as in the previous section, the physical input-output relations in (7.3.10) and (7.3.9) become respectively

$$^J Q_i = \sum_{m=1}^{m} {}^J q_{im} {}^J Q_m + \sum_{e=1}^{e} {}^J q_{ie} {}^J Q_e + {}^J Q_{iy} \qquad (7.3.23)$$

and

$$^JP_i\,^JQ_i = \sum_{m=1}^{m} {^JP_m}\,^Jq_{mi}\,^JQ_i + \sum_{e=1}^{e} {^JP_e}\,^Jq_{ei}\,^JQ_i$$

$$+ {^JP_L}\,^Jq_{Li}\,^JQ_i + {^JP_K}\,^Jq_{Ki}\,^JQ_i \tag{7.3.24}$$

Now we assume that price and quantity measures can be found such that

$$^JP_m\,^Jq_{Mi} = \sum_{m=1}^{m} {^JP_m}\,^Jq_{mi} \quad \text{and} \quad {^JP_E}\,^Jq_{Ei} = \sum_{e=1}^{e} {^JP_e}\,^Jq_{ei}$$

where $^Jq_{Mi}$ = physical quantity of the aggregate material input M required per unit output of sector i in region J, and

$^Jq_{Ei}$ = physical quantity of the aggregate energy input E required per unit output of sector i in region J.

Equation (7.3.24) can then be rewritten as

$$^JP_i\,^JQ_i = {^JP_M}\,^Jq_{Mi}\,^JQ_i + {^JP_E}\,^Jq_{Ei}\,^JQ_i$$

$$+ {^JP_L}\,^Jq_{Li}\,^JQ_i + {^JP_K}\,^Jq_{Ki}\,^JQ_i \tag{7.3.25}$$

and it is clear that the input-output coefficients $^Jq_{ri}$ (r=K,L,E,M) in (7.3.25) are identical to the cost minimizing derived factor demand per unit output $^Jq_{ri}$ (r=K,L,E,M) resulting from the use of a Diewert form for the unit cost function in the factor demand model of the previous section. To see this more clearly, recall from equation (7.2.8), (7.2.9), and (7.3.7) that

$$^Jq_{ri} = {^JQ_{ri}}/{^JQ_i} = {^Jq_{ri}} \tag{7.3.26}$$

Further, one can convert the factor demand levels $^Jq_{ri}$ (r=K,L,E,M) into the corresponding monetary unit input-output coefficients $^Ja_{ri}$ (r=K,L,E,M) by the use of equations (7.3.13), (7.3.14) and (7.3.26). That is

$$^Ja_{ri} = (^JP_r/^JP_j)\,^Jq_{ri} = (^JP_r/^JP_j)\,^Jq_{ri}$$

7.4 Linkage of Factor Demand and Input-Output Models

7.4.1 Case of Aggregated MELK Model

If the economy of region J comprised just one material input sector M and one energy input sector E, then the integration of input-output and factor demand modules could be relatively easily achieved.

a. Use of Translog Cost Function

When a translog cost function is employed, the combined operation of the factor demand and input-output modules would involve the following steps:

<u>Step 1</u>: *Estimation of Coefficients in Translog Share Equations*

Given past data on factor prices ^JP_r (r=K,L,E,M) and cost shares $^JS_{ir}$ (i=M,E; r=K,L,E,M), the coefficients $^J\alpha_{ri}$ and $^J\beta_{rs,i}$ in the translog share equations are estimated using econometric techniques.

<u>Step 2</u>: *Projection of Cost Shares Given Exogenous Factor Prices*

Given exogenous projections of factor prices for year Ω, the coefficients estimated in Step 1 are used to project cost shares for that year. That is:

$$^JS_{ri\Omega} = {}^J\alpha_{ri} + \sum_{r,s} {}^J\beta_{rs,i}\,\ln\,^JP_{s\Omega} \qquad (r,s=K,L,E,M;\ i=M,E)$$

(7.4.1)

Step 3: Revision of Input-Output Coefficients

Since the cost shares derived from equation (7.4.1) are, by definition, identical to the input-output coefficients required for year Ω, this step involves the substitution of these projected cost shares for the base year coefficients. That is, ${}^J a_{ri\Omega} = {}^J S_{ri\Omega}$ (r=K,L,E,M; i=M,E).

Step 4: Projection of Output Levels Given Exogenous Final Demands

Given exogenous projections of final demand levels for year Ω, the input-output coefficients derived in Step 3 are used to project output levels in year Ω for the endogenous producing sectors. That is, ${}^J X_\Omega = [I - {}^J A_\Omega]^{-1} {}^J Y_\Omega$ where (omitting time subscripts) we have:

$$ {}^J X = \frac{{}^J X_M}{{}^J X_E} \; ; \; {}^J A = \frac{{}^J a_{MM} \quad {}^J a_{ME}}{{}^J a_{EM} \quad {}^J a_{EE}} \; ; \; {}^J Y = \frac{{}^J X_{My}}{{}^J X_{Ey}} $$

Step 5: Projection of Labour and Capital Requirements

Given (1) the output projections derived in Step 4 and (2) exogenous projections of capital and labour necessary to satisfy final demand directly, the input-output coefficients derived in Step 3 are then used to project requirements for labour and capital inputs in year Ω. That is (omitting time subscripts):

$$ \frac{{}^J X_L}{{}^J X_K} = \frac{{}^J a_{LM}}{{}^J a_{KM}} \frac{{}^J a_{LE}}{{}^J a_{KE}} \cdot \frac{{}^J X_M}{{}^J X_E} + \frac{{}^J X_{Ly}}{{}^J X_{Ky}} $$

b. Use of Diewert Cost Function

When a Diewert (generalized Leontief) cost function is employed, the combined operation of the factor demand and input-output models would involve the following steps:

Step 1: *Estimation of Coefficients in Diewert Factor Demand Equations*

Given past data on factor price $^J P_r$ (r=K,L,E,M) and factor inputs per unit output $^J q_{ri}$ (i=M,E,; r=K,L,E,M), the coefficients $\gamma_{rs,i}$ in the Diewert factor demand equations are estimated using econometric techniques.

Step 2: *Projection of Factor Demands per Unit Output Given Exogenous Factor Prices*

Given exogeneous projections of factor prices for year Ω, the coefficients estimated in Step 1 are used to project unit factor demands for that year. That is:

$$^J q_{ri\Omega} = \sum_s \gamma_{rs,i} \, (^J P_{r\Omega} / {}^J P_{s\Omega})^{1/2} \qquad (r,s=K,L,E,M; \; i=M,E) \tag{7.4.2}$$

Step 3: *Revision of Input-Output Coefficients*

Since the $^J q_{ri\Omega}$ derived from equation (7.4.2) are, by definition, identical to input-output coefficients expressed in physical units this step involves the use of the relative price ratios $(^J P_{r\Omega} / {}^J P_{i\Omega})$ to convert these to the required input-output coefficients expressed in monetary units. That is, $^J a_{ri\Omega} = (^J P_{r\Omega} / {}^J P_{i\Omega})\, {}^J q_{ri\Omega}$ (r=K,L,E,M; i=M,E).

Step 4: *Projection of Output Levels Given Exogenous Final Demands*

As in Step 4 in (a) above.

Step 5: *Projection of Labour and Capital Requirements*

As in Step 5 in (a) above.

7.4.2 Case of Disaggregated $M(m_1, m_2, ...)$, $E(e_1, e_2, ...)$, L,K Model

Typically the economy of any region J is composed of more than one material input sector, and more than one energy sector. However, by partitioning the endogenous producing sectors (i=1,...,n) into two sets, namely those producing material inputs (m=1,...,m) and those producing energy inputs (e=1,...,e) a meaningful integration of input-output and factor demand models can still be achieved.[23].

a. Use of Across-the-Board Coefficient Adjustment

One approach would retain the use of the K,L,E and M aggregates within the factor demand module, but adopt a much larger degree of sectoral disaggregation for the input-output module. When this is done, the combined operation of the factor demand and input-output modules might involve the following steps:

Step 1: *Estimation of Coefficients in Translog Share Equations*

Given past data on factor prices $^J P_r$ (r=K,L,E,M) and cost shares $^J S_{ri}$ (i=1,...,n; r=K,L,E,M), the coefficients $^J \alpha_{ri}$ and $^J \beta_{rs,i}$ in the translog share equations are estimated using econometric techniques.

Step 2: *Projection of Cost Shares Given Exogenous Factor Prices*

Given exogenous projections of factor prices for year Ω, the coefficients estimated in Step 1 are used to project cost shares for that year. That is:

$$^J S_{ri\Omega} = {^J}a_{ri} + \sum {^J}\beta_{ri,i} \ln {^J}P_{s\Omega} \quad (r,s=K,L,E,M;\ i=1,...,n)$$

(7.4.3)

Step 3: *Revision of Input-Output Coefficients*

As in the previous section, the labour and capital cost shares derived from equation (7.4.3) are, by definition, equivalent to the corresponding input-output coefficients required for year Ω. That is, $^J a_{Li\Omega} = {^J}S_{Li\Omega}$ and $^J a_{Ki\Omega} = {^J}S_{Ki\Omega}$ for each endogenous producing sector i=1,...,n. However, the same is not true for the material and energy cost shares derived from equation (7.4.3). That is fron the cost shares $^J S_{Mi\Omega}$ and

$J_{S_{Ei\Omega}}$ we can derive projections of $J_{a_{Mi\Omega}}$ and $J_{a_{Ei\Omega}}$ for each sector i (i=1,...,n) -- yet operation of the input-output module requires projections of $J_{a_{mi\Omega}}$ (m=1,...,m) and $J_{a_{ei\Omega}}$ (e=1,...,e). However, since the following identities must hold for the latter projections:

$$ {}^J a_{Mi\Omega} = \sum_{m=1}^{m} {}^J a_{mi\Omega} \quad \text{and} \quad {}^J a_{Ei\Omega} = \sum_{e=1}^{e} {}^J a_{ei\Omega} \qquad (7.4.4)$$

the results of the factor demand module nevertheless provide a useful set of control totals to be employed when revising the input-output coefficients. For example, we could assume that coefficients relating to individual material input sectors (m=1,...,m) change by the same proportion as that projected by the factor demand module for the aggregate M of all material input sectors. That is:

$$ J_{a_{mi\Omega}} = (J_{a_{Mi\Omega}}/J_{a_{Mi\theta}}) J_{a_{mi\theta}} \qquad (7.4.5)$$

where $J_{a_{mi\theta}}$ and $J_{a_{Mi\theta}}$ represent base-year input-output coefficients (in monetary units) for the material input sector m and material input aggregate M, respectively.

Similarly, the coefficients relating to individual energy sectors (e=1,...,e) could be projected for year Ω as:

$$ J_{a_{ei\Omega}} = (J_{a_{Ei\Omega}}/J_{a_{Ei\theta}}) J_{a_{ei\theta}} \qquad (7.4.6)$$

where $J_{a_{ei\theta}}$ and $J_{a_{Ei\theta}}$ represent base-year input-output coefficients (in monetary units) for the energy input sector e and energy input aggregate E, respectively.

FACTOR DEMAND MODULE

Step 4: *Projection of Output Levels Given Exogenous Final Demands*

Given exogenous projections of final demand levels for year Ω, the input-output coefficients derived in Step 3 are used to project output levels in year Ω for the endogenous producing sectors. That is, $^J X_\Omega = [I - ^J A_\Omega]^{-1}\, ^J Y_\Omega$ where (omitting time subscripts) we have:

$$^J X = \frac{^J X_M}{^J X_E} \; ; \quad ^J A = \frac{^J A_{MM} \quad ^J A_{ME}}{^J A_{EM} \quad ^J A_{EE}} \; ; \quad ^J Y = \frac{^J X_{MY}}{^J X_{EY}}$$

where $^J X_M$ = a $(m \times 1)$ output vector with typical element $^J x_{m\Omega}$; $^J X_E$ = a output $(e \times 1)$ vector with typical element $^J x_{e\Omega}$; $^J A_{MM}$ = a $(m \times m)$ submatrix with typical element $^J a_{mm\Omega}$; $^J A_{EM}$ = a $(e \times m)$ submatrix with typical elements $^J a_{em\Omega}$; $^J A_{ME}$ = a $(m \times e)$ submatrix with typical element $^J a_{me\Omega}$; $^J A_{EE}$ = a $(e \times e)$ submatrix with typical element $^J a_{ee\Omega}$; $^J X_{MY}$ = a $(m \times 1)$ final demand vector with typical element $^J x_{my\Omega}$; and $^J X_{EY}$ = a $(e \times 1)$ final demand vector with typical element $^J x_{ey\Omega}$.

Step 5: *Projection of Labour and Capital Requirements*

Given (1) the output projections derived at the end of Step 4, and (2) exogenous projections of labour and capital necessary to satisfy final demands directly, the input-output coefficients can be used to project the requirements for labour and capital in year Ω. That is (omitting time subscripts):

$$\frac{^J X_L}{^J X_k} = \frac{^J a_{LM} \quad ^J a_{LE}}{^J a_{KM} \quad ^J a_{KE}} \cdot \frac{^J X_M}{^J X_E} + \frac{^J X_{Ly}}{^J X_{Ky}} \qquad (7.6.7)$$

where $^J A_{LM}$ = a $(1 \times m)$ labour requirements vector with typical element $^J a_{Lm\Omega}$; $^J A_{LE}$ = a $(1 \times e)$ labor requirements vector with typical element

$^J a_{Le\Omega}$; $^J A_{KM} =$ a :(1 x m) capital requirements vector with typical element $^J a_{Km\Omega}$; $^J A_{KE} =$ a (1 x e) capital requirements vector with typical element $^J a_{Ke\Omega}$.

b. Use of Restricted Coefficient Adjustment

Many objections can be raised against the simple across-the-board adjustment employed in Step 3 above to derive:

(1) projections of $^J a_{mi\Omega}$ (m=1,...,m) from projections of $^J a_{Mi\Omega}$, and
(2) projections of $^J a_{ei\Omega}$ (e=1,...,e) from projections of $^J a_{Ei\Omega}$.

Perhaps the most important of these objections is that such may result in unreasonable coefficient changes for some individual endogenous producing sectors -- that is, patterns of material and/or energy input usage that violate basic technological relations. This suggests that restrictions may need to be placed on the changes which can be projected for those sectors in which there is little or no room for adjustment in the quantity of certain inputs employed per unit output.

For example, for some sectors i, material inputs m and energy inputs e we may set

$$^J q_{mi\Omega} = {}^J q_{mi\theta} \text{ and } {}^J q_{ei\Omega} = {}^J q_{ei\theta} \qquad (7.4.8)$$

We can then derive the required monetary unit input-output coefficients for year Ω by the use of equations (7.3.12) and (7.4.8) as:

$$^J a_{mi\Omega} = ({}^J P_{m\Omega}/{}^J P_{i\Omega}) {}^J q_{mi\Omega} \text{ and } {}^J a_{ei\Omega} = ({}^J P_{e\Omega}/{}^J P_{i\Omega}) {}^J q_{ei\Omega} \qquad (7.4.9)$$

This, of course, requires projections $^J P_{m\Omega}$, $^J P_{e\Omega}$ and $^J P_{i\Omega}$. The individual material input prices $^J P_{m\Omega}$ and energy input prices $^J P_{e\Omega}$ may be either: (1) assumed to change in the same proportion as the corresponding aggregate. That is:

$$^J P_{m\Omega} = (^J P_{M\Omega} / ^J P_{M\theta}) \, ^J P_{m\theta} \quad \text{and} \quad ^J P_{e\Omega} = (^J P_{E\Omega} / ^J P_{E\theta}) \, ^J P_{e\theta}$$

(7.4.10)

or (2) derived from exogenous projections -- making sure wherever possible that these projections are not inconsistent with those for the corresponding aggregate.[24]

The sectoral output prices $^J P_{i\Omega}$ can be either derived from exogenous projections, or from the factor input prices $^J P_{M\Omega}$, $^J P_{E\Omega}$, $^J P_{L\Omega}$, and $^J P_{K\Omega}$ via the use of an appropriate price index.[25]

We can then derive the remaining input-output coefficients for year Ω by the use of a restricted across-the-board adjustment. That is

$$^J a_{mi\Omega} = (^J a_{Mi\Omega} / ^J a_{Mi\theta}) \, ^J a_{mi\theta}$$

and

$$^J a_{ei\Omega} = (^J a_{Ei\Omega} / ^J a_{Ei\theta}) \, ^J a_{ei\theta}$$

where $^J a_{mi}$ = input-output coefficients for those sectors i and material inputs m upon which no restrictions are placed concerning changes between the base year θ and projection year Ω;

$^J a_{ei}$ = input-output coefficients for those sectors i and energy inputs e upon which no restrictions are placed concerning changes between years θ and Ω; and

$^J a_{Mi} = [^J a_{Mi} - \Sigma_m \, ^J a_{mi\Omega}]$ and $^J a_{Ei} = [^J a_{Ei} - \Sigma_e \, ^J a_{ie\Omega}]$ are input-output coefficients for the aggregates of materials and energy inputs upon which no change restrictions have been placed.

Of course, further refinements can be made in terms of the nature of the restrictions placed on selected coefficients. For example, upper limits may be placed on percentage changes in some physical input-output coefficients. If the change in the corresponding aggregate coefficient falls below this limit then these input-output coefficients can be included in the across-the-board adjustments. Otherwise, these coefficients are assumed to change by the percentage specified as their upper limit. However, regardless of the restrictions made it must be that the conditions in (7.4.4) hold.

c. Use of Two Level Optimization

The analyst may be willing to assume that:

(1) the material inputs aggregate M is homothetic in its components -- that is, that the marginal rate of substitution between materials within this aggregate depend only on the prices P_m (m=1,...,m) of these materials; and

(2) the energy inputs aggregate E is homothetic in its components -- that is, that the marginal rate of substitution between energy inputs within this aggregate depend only on the prices P_e (e=1,...,e) of these energy inputs.

That is, he may be willing to assume that the composition of the material and energy inputs aggregates, respectively, are independent of the results of the overall substitution between capital, labour, energy and materials.

If this is so, then a two-stage optimization procedure may be employed within the factor demand module.[26] That is, the optimal mix of inputs within each aggregate is determined first; and then the optimal level of each aggregate. More specifically, we may derive the share $^JS_{mi}$ of material input m (m=1,...,m) in the aggregate material costs per unit output as:[27]

$$^JS_{mi} = \delta_{mi} + \sum_{m'=1}^{m} \gamma_{mm',i} \ln{^JP_{m'}}$$

(7.4.11)

where $\sum_m \delta_{mi} = 1$; $\sum_m \gamma_{mm',i} = \sum_{m'} \gamma_{mm',i} = 0$; and $\gamma_{mm',i} = \gamma_{m'm,i}$; m ≠ m'.

Similarly, we may derive the share $^JS_{ei}$ of energy input e (e=1,...,e) in the aggregate energy costs per unit output as:[28]

$$^J S_{ei} = \delta_{ei} + \sum_{e'=1}^{e} \gamma_{ee'i} \ln\, ^J P_{e'} \qquad (7.4.12)$$

where $\Sigma_e\, \delta_{ei} = 1$; $\Sigma_e\, \gamma_{ee'i} = \Sigma_{e'}\, \gamma_{ee,i} = 0$; and $\gamma_{ee',i} = \gamma_{e'e,i}$, $e \neq e'$.
Then employing: (1) the $^J S_{mi}$ ($m=1,...,m$) and $^J P_m$ to derive the price $^J P_M$ for the materials aggregate M, and
(2) the $^J S_{ei}$ ($e=1,...,e$) and $^J P_e$ to derive the price $^J P_E$ for the enregy aggregate E,[29] the analyst can then determine the unit cost shares for the K,L,E, and M aggregates using equation (7.2.4) for each sector i ($i=1,...,n$) in region J.

When this two-staged optimisation procedure is employed, the combined operation of the factor demand and input-output modules would involve the following steps:

Step 1: *Estimation of Coefficients in Translog Share Equations*
  Given past data on (1) factor prices $^J P_r$ ($r=K,L,E,M$), $^J P_m$ ($m=1,...,m$) and $^J P_e$ ($e=1,...,e$) and (2) cost shares $^J S_{ri}$, $^J S_{mi}$, and $^J S_{ei}$ ($i=1,...,n$), the coefficients in the stochastic versions of the translog share equations (7.4.11), (7.4.12) and (7.2.4) are estimated using econometric techniques.

Step 2: *Projection of Cost Shares Given Exogenous Factor Prices*
  Given exogenous projections of factor prices $^J P_{m\Omega}$ ($m=1,...,m$) and $^J P_{e\Omega}$ ($e=1,...,e$), $^J P_{L\Omega}$ and $^J P_{K\Omega}$ for year Ω, the coefficients estimated in Step 1 are used to project cost shares $^J S_{mi\Omega}$ and $^J S_{ei\Omega}$ for individual material and energy inputs. These factor price and cost share projections are then used to calculate the corresponding aggregate material and energy prices, $^J P_{M\Omega}$ and $^J P_{E\Omega}$, respectively. Then, these latter prices are used together with the coefficients estimated in Step 1 to project aggregate cost shares $^J S_{ri\Omega}$ ($r=K,L,E,M$).

Step 3: *Revision of Input-Output Coefficients*

The labour and capital cost shares derived from Step 2 are, by definition, equivalent to the corresponding input-output coefficients required for year Ω. That is, $^J a_{Li\Omega} = {}^J S_{Li\Omega}$ and $^J a_{Ki\Omega} = {}^J S_{Ki\Omega}$ for each endogenous producing sector (i=1,...,n).

The material and energy cost shares derived from Step 2 can also be used to revise the corresponding input-output coefficients required for year Ω. That is $^J a_{mi\Omega} = ({}^J S_{mi\Omega})({}^J S_{Mi\Omega})$ and $^J a_{ei\Omega} = ({}^J S_{ei\Omega})({}^J S_{Ei\Omega})$ for each endogenous producing sector (i=1,...,n).

Step 4: *Projection of Output Levels Given Exogenous Final Demands*

As in Step 4 of part (a) of this section.

Step 5: *Projection of Labour and Capital Requirements*

As in Step 5 of part (a) of this section.

7.5 Integration of Multiregional Econometric, Factor Demand, and Input-Output Modules

Step 1: *Identification of Sectoral Structure and Setting Up of Modified Interregional Input-Output Model*

Each region's economy is divided into four sets of sectors: material input sectors M, energy input sectors E, labour input sectors L and capital input sectors K. The base year input-output model is partitioned accordingly.

Step 2: *Derivation of Initial Final Demand Projections*

An initial set of final demand vectors $^J Y' = [{}^J Y'_M \ {}^J Y'_E \ {}^J Y'_L \ {}^J Y'_K] = [{}^J X'_{MY} \ {}^J X'_{EY} \ {}^J X'_{LY} \ {}^J X'_{KY}]$ is projected for year t+θ. This can be done either:

(a) by allocating national projections of these magnitudes among regions using the best allocators available; or

(b) by use of trend analysis or other "bottom-up" approaches.

Step 3: *Derivation of Initial Output Projections*

Given the final demand vector derived in Step 2, the interregional input-output model can be run in standard manner -- using base year technical-trade coefficients -- to yield a projection of output levels by sector by region in the year t+θ. That is: $X_{t+\theta} = [I-CA]^{-1} Y_{t+\theta}$

Step 4: *Derivation of Initial Employment Demand Projections*

Given the output projections derived in Step 3, labour requirements (employment demand) by sector can be projected via the use of a set of "reasonable" output to employment coefficients (J_{w_i} and $J_{w_{id}}$)[30] as in equations (6.2.17) and (6.2.18).

Step 5: *Derivation of Initial Population Projections*

Given the regional employment demand projections derived in Step 4, population by region can be projected as in Step 4 of Section 6.3.1.

Step 6: *Derivation of Initial Labour Supply Projections*

Using the results of Step 5 (population level projections), as in Step 5 of Section 6.3.3.

Step 7: *Derivation of Regional Wage Rate and Wage Income Projections*

Using the results of Steps 4 (regional employment demands) and 6 (regional labour supplies), as in Steps 6 and 7 of Section 6.3.3.

Step 8: *Derivation of Regional Household Disposable Income Projections*

Using the results of Steps 5 (population level projections) and 7 (regional wage and salary incomes), as in Step 6 of Section 6.3.1.

Step 9: Derivation of Regional Household Consumption Expenditure Projections

Using the results of Steps 5 (population level projections) and 8 (household disposable income), as in Steps 7 and 8 of Section 6.3.1.

Step 10: Derivation of Regional Government Expenditure Projections

Using the results of Steps 5 (population level projections), 8 (household disposable income) and 3 (output by sector), as in Steps 9 and 10 of Section 6.3.2.

Step 11: Derivation of Regional Private Investment Expenditure Projections

Using the results of Steps 3 (output by sector), 5 (population level projections), 7 (regional wage rates and wage and salary incomes), as in Steps 13 and 14 of Section 6.3.4.

Step 12: Revision of Regional Final Demand Projections

The household consumption expenditure projections derived at the end of Step 9, the government expenditure projections derived at the end of Step 10, and the private investment expenditure projections derived at the end of Step 11 are used to revise the initial final demand projections made in Step 2.

Step 13: Projection of Regional Cost Shares for Each Factor of Production

Given past data on factor prices and cost shares, the coefficients in the translog cost share equations comprising the factor demand module are estimated using econometric techniques. Then given the wage rates JP_{Li} derived in Step 7 and exogenous projections of other factor prices (JP_{Mi}, JP_{Ei} and JP_{Ki}) for year $t+\theta$, these estimated coefficients are used to project regional cost shares for each factor of production.

FACTOR DEMAND MODULE 281

Step 14: *Revision of Interregional Input-Output Coefficients*

The base year interregional technical input-output coefficients employed in Step 3 are revised for year t+θ using the cost shares projected in Step 13, and any restrictions which the analyst chooses to impose.[31]

Step 15: *Revisions of Regional Output Projections*

Given the revised final demand vector derived in Step 12 and the revised input-output coefficients matrix derived in Step 14, the interregional input-output model is rerun to yield a revised projection of output levels by sector by region.

Steps 4-15 are then repeated as many times as required for the integrated model to converge on a reasonable set of results.

7.6 Integration of Multiregional Demographic, Factor Demand and Input-Output Modules

Step 1: *Derivation of Initial Population Projections*

As in Steps 1 and 2 of Section 5.4.

Step 2: *Identification of Sectoral Structure and Setting up of Modified Interregional Input-Output Model*

As in Step 1 of Section 7.5.

Step 3: *Derivation of Initial Final Demand Projections*

As in Step 2 of Section 7.5.

Step 4: *Projection of Regional Cost Shares for Each Factor of Production*

As in Step 13 of Section 7.5, except that regional wage rate projections $^J P_{Li}$ for year t+θ must now be taken as exogenous.

Step 5: *Revision of Base-Year Interregional Input-Output Coefficients*

The technical component of the base-year interregional input-output coefficients identified in Step 2 are now revised for year $t+\theta$ using the cost shares projected in Step 4, and any restrictions which the analyst chooses to impose.[32]

Step 6: *Derivation of Initial Output Projections*

Given the final demand vector derived in Step 3 and the interregional input-output coefficients derived in Step 5, the interregional input-output model can be run in standard manner to yield a projection of output levels by sector by region.

Step 7: *Derivation of Initial Labour Requirements Projections*

The next step involves the use of the output estimates derived in Step 6 to get an initial estimate of employment demand (labour requirements) in each region for year $t+\theta$. This is done by the use of a set of "reasonable" output to employment coefficients for each sector in each region, taking into consideration the ${}^J a_{Li}$ ($i=1,...,n$) coefficients derived in Step 5. These former coefficients can relate to total labour requirements or be broken down by demographic class depending on the level of disaggregation required for the operation of the migration component of the demographic model.

Step 8: *Derivation of Crude Wage and Salary Income Projections*

The next step involves the derivation of crude estimates of regional wage and salary incomes from the year $t+\theta$. Where the migration component of the demographic model does not require demographic class specific wage and salary income projections, then this is done crudely via the use of standard input-output procedures. That is, each sector's output level ${}^J X_i$ derived at the end of Step 6 is multiplied by the corresponding ${}^J a_{Li}$ coefficient derived in Step 5. The resulting wage and salary incomes are then summed over all sectors -- that is $^J WY = \Sigma_i\, ^J a_{Li}\, ^J X_i$.

However, when regional wage and salary income projections are required for each demographic class then a set of demographic class specific wage rates must be specified for each region -- making sure they are not inconsistent with the corresponding $^J P_L$ used in Step 4. These wage rates are then multiplied by the employment demand projections by demographic class derived in Step 7, to give the required set of wages and salary income projections. These projections are then summed over all demographic classes to yield a total wage and salary income projection $^J WY'$ for each region. Should major inconsistencies arise between the $^J WY$ and $^J WY'$ projections, they are eliminated through the use of a series of proportional adjustments.

Step 9: *Derivation of Initial Labour Supply Projections*

The next step involves the use of the population estimates derived at the end of Step 2 to get initial crude regional labour force projections (excluding the effects of economically-induced interregional migration flows) for year t+θ. This is done via the use of a set of relevant (crudely estimated) labour force participation rates for each region.

Step 10-12: *Derivation of Gross Outmigration and Inmigration Rate Projections*

Using the results of Step 7 (regional employment demands), Step 8 (regional wage and salary incomes) and Step 9 (regional labour supplies), as in Steps 9-11 of Section 5.4.

Steps 13-15: *Derivation of Net Interregional Migration Flow Projections*

Using the results of Step 12 (gross regional migration levels), as in Steps 12-14 of Section 5.4. These estimates of the economically induced change in each region's population over the projection period are then used to revise the initial crude population projections derived for each region at the end of Step 2.

Step 16: *Revision of Final Demand Projections*
The population estimates derived at the end of Step 15 are then used to revise the initial final demand projections made in Step 4.

Steps 6-16 can then be repeated as many times as is required in order to converge on a reasonably consistent set of results.

7.7 Integration of Multiregional Input-Output Comparative Cost-Industrial Complex, Programming, Demographic, Econometric and Factor Demand Modules

Step 1: *Derivation of Crude Population Projections*
The first step involves the derivation of crude estimates of population for each region at the end of the projection period. Excluding the effects of foreign and interregional migration, these crude estimates are found by aging the corresponding base year populations in each year. This involves estimating the number of births and deaths during the projection period for this base year population, as well as adjusting for non-interregional movements from one demographic class to another during this period. See Step 1 of Section 5.4 for further details.

Step 2: *Adjustment for Non-Economically Induced Migration*
The second step involves adjustment of the crude regional population estimates derived in Step 1 for the levels of those types of population movements that are assumed not to be directly dependent on prevailing regional and/or national economic conditions -- that is, for the level of net foreign immigration and net interregional migration of persons aged 65 and over. See Step 2 of Section 5.4 for further details.

Step 3: *Identification of Sectoral Structure and Setting Up of Interregional Input-Output Model*

Each region's economy is divided into six sets of sectors: (i) energy product sectors P, (ii) energy supply sectors S, (iii) cost sensitive non-energy sectors N', (iv) other non-energy sectors N", (v) labour input sectors L, and (vi) capital input sectors K. The cost sensitive non-energy sectors N' are removed from the interindustry section of the standard input-output transactions table, and placed within a set of "non-traditional" final demand columns/primary input rows.[33] The modified base year interregional technical-trade coefficients matrix can then be partitioned in the manner required for the joint operation of the factor demand, programming and input-output modules.[34]

Step 4: *Derivation of Initial Traditional Final Demand Projections*

The next step involves the derivation of an initial set of traditional final demand vectors $^JY' = [^JY'_{N"} \ ^JY'_S \ ^JY'_P \ ^JY'_{N'} \ ^JY'_L \ ^JY'_K]$ for the year t+θ. This can be done using the results of Step 1 (regional population levels either, (a) by allocating national projections of these magnitudes among regions using the best allocators available; or (b) by use of trend analysis or other "bottom-up" approaches.[35]

Step 5: *Derivation of Initial Market Projections for Cost Sensitive Sectors*

The next step involves the estimation of the level of regional demands (intermediate and final) for the output of cost sensitive non-energy sectors in the year t+θ.[36] (The final demand projections $Y_{N"}$ were estimated in Step 4, so essentially this step involves projection of the requirements for these sectors' outputs as intermediates).

Step 6: *Derivation of Initial Non-Tradtional Final Demand Projections*

The next step involves the use of the multiregional variants of the comparative cost and/or industrial complex approaches to derive projections for each region of output levels and required input-output coefficient changes for cost-sensitive non-energy sectors.[37]

Step 7: *Consolidation of Traditional and Non-Traditional Final Demand Projections*

Now that the level of "traditional" and "non-traditional" final demand by sector by region have been projected for year $t+\theta$ (in Steps 4 and 6, respectively), they can now be consolidated into the combined traditional and non-traditional final demand vectors $Y'_{N''}$, Y'_S, Y'_P, and $Y'_{N'}$, required for operation of the input-output and programming modules.

Step 8: *Derivation of Initial Output Projections for Non-Energy and Energy Product Sectors*

A first approximation to the output requirements for the other non-energy sectors N" and energy product sectors P can then be calculated respectively as:

$$X_{N''} = [I - \mathcal{Æ}_{N''N''}]^{-1} Y_{N''}$$

and

$$X_P = \mathcal{Æ}_{PN''} X_{N''} + Y_P$$

Step 9: *Projection of Functional Energy Service Demand Levels*

The output estimates X_P derived in Step 8 are used to provide a first approximation of regional demands for energy end-products. By employing a set of aggregation and/or disaggregation procedures, these estimates can then be converted into the functional energy service demand levels ^JD_d (d=1,...,17) required as binding magnitudes for the demand equations within the linear programming module.

Step 10: *Derivation of Initial Output Projections for Energy Supply Sectors*

The linear programming module is now run to yield a first approximation of the output requirements X_S of the energy; supply sectors -- as well as the $Æ_{SS}$ and $Æ_{SP}$ submatrices.

Step 11: *Successive Reruns of Input-Output and Programming Modules*

The results of Step 10 are now used to derive revised estimates of $X_N"$, X_P, and ^JD_d (J=A,...,U; d=1,...,17). The linear programming module is rerun, and new projections of X_S derived. And so on, with Steps 8-11 repeated until a point is reached where the input-output and programming modules converge on a stable set of outputs for the energy product P, energy supply S, and other non-energy N" sectors.[38]

Step 12: *Identification of Inconsistencies in Output Projections*

The set of output vectors X_N', $X_N"$, X_P and X_S derived from a firstrun through of Steps 1-11 can be expected to contain a number of inconsistencies. For example:[39]

(a) for cost sensitive non-energy industries, between the pattern of regional outputs derived at the end of Steps 6 and 11 and the pattern of regional markets estimated in Step 5;

(b) for cost sensitive non-energy industries, between regional cost differentials (especially re: energy) calculated in Step 6 and the results of Step 11; and

(c) for each region, the output vectors X_N' (derived at the end of Step 6) and $X_N"$, X_P, X_S (derived at the end of Step 11) may be inconsistent with the level of traditional final demand by type by sector estimated for that region at the end of Step 4.

Step 13: *Elimination of Major Inconsistencies*

If any of the inconsistencies in Step 12 are judged to be too large to be permitted in the final projections, the analyst must re-estimate one or more of:

(a) the level of traditional final demand by sector by region;

(b) the regional pattern of industrial markets (intermediate demands) for the output of cost-sensitive sectors; or

(c) the regional cost differentials associated with one or more of the cost sensitive sectors.

The results of Step 10 provide the basic materials required to perform these re-estimations in (b) and (c); while re-estimation of the level of traditional final demand by sector by region can be performed using steps 14-31 below.

Step 14: *Derivation of Initial Employment Demand Projections*

Given the output projections derived in Step 11, as in Step 4 of Section 7.5.

Step 15: *Derivation of Initial Labour Supply Projections*

Given the population projections derived in Step 2, labour supply can be projected for each region and demographic class via use of the econometric equations (6.2.20)-(6.2.24) of the labour market submodule discussed in Section 6.2.2. Note that such excludes the effects of economically-induced interregional migration flows.

Step 16: *Derivation of Regional Wage Rate Projections*

Given the employment demand projections derived in Step 14 and the labour supply projections derived in Step 15, initial crude regional wage rate projections can be derived for each sector via use of the econometric equations (6.2.25)-(6.2.29) of the labour market submodule discussed in Section 6.2.1.

Step 17: *Derivation of Regional Wage Income Projections*

Regional wage income projections can then be made via use of the identity in equation (6.2.1) of the household income and consumption expenditure submodule -- that is, by multiplying the employment demand projections derived in Step 14 by the regional wage rate projections derived in Step 16, and then summing overall sectors (and demographic classes).

Step 18: *Projection of Gross Outmigration and Gross Inmigration Rates for Each Region's Labour Force*

The results of Step 14 (regional employment demands), Step 15 (regional labour supplies) and Step 17 (regional wage and salary incomes) can than be used as RHS variables in the migration equations (5.3.32) and (5.3.33) discussed in Section 5.3.1. This yields projections of gross inmigration and outmigration rates for each region's labour force over the projection period.

Step 19: *Estimation of Labour Force "At Risk" of Migration Over Projection Period*

This is done by simply averaging the base year labour supply with that estimated in Step 15 as the labour force level at the end of the projection period.

Step 20: *Projection of Gross Labour Force Outmigration and Inmigration Levels for each Region*

Multiplication of the migration rates (derived in Step 18) by the corresponding 'at risk' labour force (derived in Step 19) yields estimates of the level of gross outmigration and inmigration for each region's labour force.

Step 21: *Projection of Net Interregional Labour Force Migration Flows*

The results of Step 20 can now be employed in the doubly constrained gravity model described in Section 5.3.1 to yield an estimate of net interregional migration flows for members of the labour force over the projection period.

Step 22: *Projection of Net Interregional Non-Labour Force Migration Flows*

As in Step 15 of Section 7.6, this step involves estimation of interregional migration flows for (a) persons aged 15-64 who are not members of the labour force, and (b) persons aged 0-14. This is done using the procedures described in Steps 13 and 14 of Section 5.4.

Step 23: *Revision of Regional Population Projections*

The results of Steps 21 and 22 can now be combined to yield an estimate of the economically induced change in each region's population over the projection period. These estimates are then used to revise the initial crude population projections derived for each region at the end of Step 2.

Step 24: *Derivation of Regional Household Disposable Income Projections*

The next step involves the use of the econometric equations (6.2.3)-(6.2-12) of the household income and consumption expenditure submodule discussed in Section 6.2.1 to derive projections for:

(a) non-wage income by type by region -- using the results of Step 11 (output by sector) and 23 (revised population levels by demographic class);

(b) household income by region -- as an identity from (a) and the results of Step 17 (wage incomes);

(c) household disposable income by region; and

FACTOR DEMAND MODULE   291

(d) household disposable income per capita by region -- as an identity from (c) and the results of Step 23 (revised population levels).

Step 25: *Derivation of Regional Household Consumption Expenditure Projections*

Using the results of Steps 23 (population by demographic class) and 24 (household disposable income) in econometric equations (6.2.13)-(6.2.16) of the household income and consumption expenditure submodule discussed in Section 6.2.1 yields projections of household consumption expenditure by type by region. These projections are then converted into estimates of household consumption expenditure by sector via use of a transformation matrix of the form discussed in Step 8 of Section 6.3.1.

Step 26: *Derivation of Regional Government Expenditure Projections*

The next step involves the use of the results of Step 23 (population by demographic class), Step 24 (household disposable income) and Step 11 (output by sector) in the econometric equations (6.2.33)-(6.2.45) of the government expenditure submodule discussed in Section 6.2.4. This yields projections of government expenditures disaggregated by type by level by region, which are then converted into government expenditure estimates disaggregated by sector via use of a transformation matrix of the form discussed in Step 10 of Section 6.3.2.

Step 27: *Derivation of Regional Private Investment Expenditure Projections*

Using the results of Steps 11 (output by sector), 16 (wage rates), 17 (wage incomes), 23 (population levels) and 24 (household disposable income) in the econometric equations (6.2.30)-(6.232) of the private investment expenditure submodule, discussed in Section 6.2.3, yields projections of private investment expenditure by sector and region in which the investment is to be carried out. These projections are then converted, via a transformation matrix of the form discussed in Step 14 of Section 6.3.4., into estimates of private investment expenditure by sector and region which supplies the goods and services required to carry out the investment.

Step 28: *Revision of Regional Traditional Final Demand Projections*
The household consumption expenditure projections derived at the end of Step 25, the government expenditure projections derived at the end of Step 26 and the private investment expenditure projections derived at the end of Step 27 are used to revise the initial traditional final demand projections made in Step 4.

Step 29: *Projection of Cost Shares for Non-Energy Sectors*[40]
Given past data on factor prices ^JP_P, $^JP_{N'}$, $^JP_{N''}$, ^JP_L, and ^JP_K and cost shares $^JS_{Pn}$, $^JS_{N'n}$, $^JS_{N''n}$, $^JS_{Ln}$ and $^JS_{Kn}$, the coefficients $^J\alpha_{rn}$ and $^J\beta_{rs,n}$ (r= N',N",P,L,K) in the translog cost share equations are estimated for each non-energy sector n using econometric techniques.[41] Then given projections of ^JP_r (r=N',N",P,L,K) for year t+θ, these estimated coefficients are used to project the corresponding cost shares for that year. That is:

$$^JS_{rn}(t+\theta) = {^J\alpha_{rn}} + \sum {^J\beta_{rs,n}} \ln {^JP_r(t+\theta)} \quad (r,s=N,P,L,K;\ n\epsilon N)$$

The ^JP_N, ^JP_L and ^JP_K projections must be derived from outside the integrated model. However, for each energy end-product sector p∈P the required price projection can be derived from the solution of the linear programming module at the end of Step 16. That is, by assuming the price of end-product p is a function of (1) the costs of extracting (including scarcity rents) of the various energy resources s entering into it; (2) the cost of converting these energy resources into the required end-product; (3) the costs of transporting these energy resources from the point of their extraction to the point of their conversion; and (4) the cost of transporting (transmitting) the energy end-product from its point of production to its point of final use in the given non-energy sector.

Step 30: *Revision of Input-Output Coefficients*

The base year regional technical coefficients $^J a_{PN'}$, $^J a_{N"N'}$, $^J a_{LN'}$, $^J a_{KN'}$, $^J a_{PN"}$, $^J a_{LN"}$, $^J a_{N'N"}$ and $^J a_{KN"}$ coefficients can now be revised for year $t+\theta$, using the cost shares projected in Step 24, and any restrictions which the analyst may choose to impose.[42] For the $^J a_{N'N"}$, $^J a_{LN"}$, $^J a_{KN"}$, $^J a_{LN'}$ and $^J a_{KN'}$ coefficients, the procedures which can be employed to achieve the necessary revisions are the same as those discussed in Section 7.4. However, for the $^J a_{PN'}$, $^J a_{PN"}$, and $^J a_{N'N"}$ coefficients a more complex set of procedures is required. The typical coefficient in the $^J a_{PN"}$ submatrix for example, is expressed in mixed physical-monetary units -- that is it specifies the physical quantity of energy product p that is required per dollar output of non-energy sector n". In mathematical terms, $^J a_{pn"} = {^J}Q_{pn"}/{^J}X_{n"}$. Providing all energy end products are measured in comparable units, say BTU's of energy, we can still make use of the identity $^J a_{Pn"} = \Sigma_p {^J}a_{pn"}$ for each n"εN". However, the cost share $^J S_{pn"}$ is no longer identical to the input-output coefficient $^J a_{pn"}$ -- rather we must make use of the relation $^J a_{pn"} \, {^J}P_p = {^J}S_{pn"}$ for each non-energy sector n"εN" before employing any of the procedures discussed in Section 7.4.

The $^J a_{pn'}$ and $^J a_{n"n'}$ coefficients for the cost sensitive non-energy sectors n"εN" are those within the non-traditional final demand vectors -- and not within the input-output structural matrix. Further, these $^J a_{pn"}$ and $^J a_{n"n'}$ coefficients are in addition converted into their physical counterparts $^J q_{pn'}$ and $^J q_{n"n'}$ for use in revised comparative cost and for industrial complex analyses.

Step 31: *Rerun of Integrated Model*

Steps 5-30 are then repeated incorporating:

(a) the revised traditional final demand projections derived at the end of Step 28;

(b) the revised industrial markets for the output of cost sensitive non-energy sectors derived at the end of Step 13;

(c) the revised cost differentials (especially re: energy) derived at the end of Step 13;

(d) the revised wage rates derived at the end of Step 16;

(e) the revised wage rates derived at the end of Step 23; and

(f) the revised input-output coefficients (expressed in monetary and/or physical units) derived at the end of Step 30.

A number of such reruns may be required until all major discrepancies are eliminated -- that is until the integrated model converges on a reasonable set of results.

7.8 Concluding Remarks

This chapter has dealt with the factor demand module and its linkage to the input-output, comparative cost-industrial complex, linear programming, demographic and econometric modules of an integrated multiregion model.

In the Australian context this module can only be made operational for manufacturing sectors, and for these sectors we must pool time series and cross sectional data in order to get enough observations for the results to be meaningful. Even applied in this restricted manner, it may be that the cost share equations prove virtually useless for projection purposes when they are confronted with actual data. If this is the case, then we suggest that the analyst either:

(1) make use of expert opinion and ad hoc adjustment to selected input-output coefficients, or

(2) identify past trends in the cost shares for capital, labour energy and materials, and assume these trends continue in the future, or

(3) adjust the trends identified in (2) on the basis of expert opinion as to the likely pattern of changes in these trends.

If one of the options given in (2) or (3) is employed, then the required input-output coefficient changes can be made using the procedures described in Section 7.4.

NOTES

[1] In doing so, it draws upon earlier work by Hudson and Jorgensen (1974, 1976), Lakshmanan (1979, 1980, 1981), Lesuis, et al (1980) and Lakshmanan, et al (1980,1981).

[2] The CES production function assumes a constant elasticity of substitution between factors of production, the Cobb Douglas function assumes a unitary elasticity of substitution, and use of a Leontief production function implies zero elasticity of substitution between factor inputs. Further, none of these functions is flexible enough to permit differences in the elasticities of substitution between pairs of inputs when more than two exist. See Henderson and Quandt (1980) for further discussion of the properties of these functional forms.

[3] See Fuss, et al (1978) for a survey of these more general functional forms. See also Christensen and Lau (1971), Christensen, et al (1973) and Diewert (1971) for the original development of the translog and generalized Leontief functional forms.

[4] See Diewert (1974) and the comments contained therein for a short, yet complete survey of duality theory. See Fuss and McFadden (1978, Vols. 1 and 2) for a comprehensive treatment of duality theory as well as numerous applications.

[5] Namely, that the input requirements set be regular, convex and monotonic (Varian, 1978).

[6] Prior to this development, the approach had been to select a particular form for the production function and then to derive the equilibrium factor demand levels from the solution of a constrained optimisation problem. While this posed no real problem in the case of simple forms such as Cobb Douglas, the analytical derivation becomes very difficult for more general and complex specifications of the production function. (Diewert, 1974; Varian, 1978).

[7] Indeed, the number of applications of this approach has been growing fairly rapidly. See Berndt and Christensen (1973); Hudson and Jorgenson (1974, 1976), Berndt and Wood (1975), Christensen and Greene (1976), Griffin and Gregory (1976), Fuss (1977), Halvorsen (1977), Halvorsen and Ford (1978), Almon, et al (1979), Emerson and Hunt (1980), Field and Grebenstein (1980), Lesuis, et al (1980), and Lakshmanan, et al (1981).

[8] In more rigorous terms, we restrict the production structure to be homothetic. Such allows the cost function to be written as a separable function in output level and factor prices. See Hudson and Jorgenson (1976) and Christensen and Greene (1976). Although the translog cost function could conceptually be applied to as many inputs as desired, in practice its application can be cumbersome when many inputs are involved in the simultaneous estimation of parameters. Also, multicollinearity and degrees of freedom problems would often arise in the absence of this restriction -- since the analyst generally has only a very limited time series of observations and lack of variation in input price and quantity data over such short time spans.

[9] In more rigorous terms, we restrict the production structure not only to be homothetic, but also to be homogeneous of degree one. This implies that the elasticity of total production costs with respect to output is constant. See Christensen and Greene (1976).

[10] This functional form has found widespread use because it imposes no a priori restrictions on the Allen elasticities of substitution, and because it can be viewed as a second order local approximation (or Taylor series expansion) in logarithms to any cost function. See Christensen, et al (1973) and Denny and Fuss (1977).

[11] That is $(\delta C_i/\delta P_r)Q_i = Q_{ri} = q_{ri}Q_i$, which simplifies to the expression given above.

[12] More rigorously these conditions are required in order for cross price effects to be symmetric and for the cost function in (7.2.2) to be homogeneous of degree one in prices -- properties required in order for that cost function to represent a well-behaved production function. See Varian (1978).

[13] The error terms μ_{ri} reflect such factors as economies of scale, technical change, 'irrationality' of producers, restrictions imposed by different levels of government, exchange rate fluctuations for firms involved in export trade, and fluctuations in interest rates reflecting 'ups' and 'downs' in the stock market. Because such a multitude of factors is involved, the assumptions of constant variances and zero means may be reasonably well aproximated.

[14] Since the system of equations in (7.2.6) is singular, one equation is deleted from the estimation process and its coefficients derived as residuals from the restrictions given in (7.2.5). The iterative Zellner procedure is then appropriate to obtain asymptotically efficient

parameter estimates equivalent to maximum likelihood methods. See Hudson and Jorgensen (1976) and Emerson and Hunt (1980).

[15]Once again, this functional form imposes no a priori restrictions on the Allen elasticities of substitution, and can be viewed as a second order local approximation to any arbitrary cost function. (Diewert, 1974). One many wonder why the term $(P_r/P_s)^{1/2}$ appears in (7.2.7) rather than the more general $(P_r/P_s)^\beta$ where ß is a parameter to be estimated. The answer is that (7.2.8) can derive from cost minimising behavior if and only if ß=1/2. This is so since symmetry of the cross partials of the demand functions implies that $\delta q_{ri}/\delta P_s = \delta q_{si}/\delta P_r$. But the corresponding condition $\beta \, \gamma_{rs,i} \, P_s^{\beta-1} \, P_r^{-\beta} \, \gamma_{sr,i} \, P_r^{\beta-1} \, P_s^{-\beta}$ holds if and only if ß=1/2. See Almon, et al (1979).

[16]Note that if $\gamma_{rs,i} = 0$ for all r ≠ s, then (7.2.9) simplifies to $Q_{ri} = \gamma_{rr,i} \, Q_i$ for all r, and we have a system of demand functions corresponding to a fixed coefficient (Leontief) production function. It is for this reason that the Diewert functional form is often designated the Generalized Leontief. See Diewert (1971, 1974).

[17]For a more detailed discussion of this potential refinement, see Christensen and Greene (1976), Lakshmanan (1979, 1980) and Lakshmanan, et al (1980, 1981).

[18]For further discussion of this potential refinement, see Almon, et al (1979) and Lakshmanan (1979, 1980, and 1981).

[19]When long-run projections are being made, this is not such a bad assumption since the relative price changes projected over the projection period (say between year t and t+θ) can be taken as having occurred gradually and in large part over that period. Hence by year t+θ the firms will have (more or less) completed their adjustment process.

[20]When a translog cost function is employed, a similar partial adjustment model could be set up to relate the desired or equilibrium cost shares to actual cost shares. See Lakshmanan (1981) for further discussion.

[21]Note that the prices in the RHS of this equation must be delivered prices in region J -- hence we have ^JP_i and ^JP_r in the first and second terms respectively.

[22] Note that the discussion in this section assumes that the final demand component of the input-output model refers to net foreign exports (or that foreign imports are allocated indirectly), so that the $^J a_{ri}$ coefficients indicate <u>technical</u> input requirements.

[23] In order to keep things simple, this section will restrict its attention to the use of a translog cost function within the factor demand module. By referring to the discussion in Section 7.4.1, the reader can develop the parallel set of steps should he choose to use a Diewert cost function.

[24] There will inevitably be some inconsistency, since the aggregate price should be a weighted average of the prices of its components -- with the weights being quantities of these components employed during year Ω. Yet, these quantities are what we are seeking to determine.

[25] See Hulten (1973) and Diewert (2976) for discussion of some indexes which may be employed. However, once again the quantities of the component inputs employed in year Ω must be used as weights in construction of these indicies. Thus, conceptually at least, a number of iterations may be required before an appropriate set of sectoral output prices can be derived.

[26] For further discussion of this two-staged optimisation process, and the conditions under which its use can be justified, see Hudson and Jorgenson (1976), Fuss (1977) Lesuis, <u>et al</u> (1977), and Lakshmanan, <u>et al</u> (1980,1981).

[27] Since the system of equations in (7.4.11) is singular, one equation is deleted from the estimation process and its coefficients derived as residuals from the set of restrictions specified above.

[28] Since the system of equations in (7.4.12) is singular, one equation is deleted from the estimation process and its coefficients derived as residuals from the set of restrictions specified above.

[29] See Hulten (1978) and Diewert (1976) for discussion of some indexes that may be used for deriving these aggregate prices.

[30] Clearly these coefficients depend on (are functionally related to) the input-output coefficients $^J a_{Li}$ and the wage rates $^J w_i$ for sector i (i=1,...,n). In later rounds then, they will need to be revised in light of results of Steps 7 and 13, discussed below.

[31] See Section 7.4.2 for discussion of the nature of some reasonable restrictions which may be imposed.

[32] See Section 7.4.2 for a discussion of the nature of some reasonable restrictions which may be imposed.

[33] See Step 3 of Section 3.3 for further details.

[34] See Section 4.3.2. for further details.

[35] See Steps 1 and 2 of Section 3.3 for further details.

[36] See Section 3.2 and Step 4 of Section 3.3 for further details.

[37] See Section 3.2 and Step 5 of Section 3.3 for further details.

[38] Steps 8-11 above resemble closely those given as Steps 2-9 in Section 4.3.2, to which the reader is referred for a more detailed discussion.

[39] These inconsistencies resemble those given on pp. 61-62, to which the reader is referred for a more detailed discussion.

[40] Thus the factor demand module is, in effect, being restricted to application for the non-energy sectors only. This restriction is also dictated by lack of the required data for non-manufacturing sectors -- which include the energy sectors.

[41] Note that we have excluded $J_{S_{S_n}}"$. This is so since, by definition, $J_{S_{S_n}}" = 0$ for all non-energy sectors $n"\varepsilon N"$.

[42] See Section 7.4.2 for a discussion of the nature of reasonable restrictions that may be imposed.

CHAPTER 8
POTENTIAL EXTENSIONS TO THE INTEGRATED MODEL

8.1 Introduction

The integrated multiregion model developed in Section 7.7 can be extended to include other non-policy modules. For example, the inclusion of a national econometric module designated NATLEC is discussed in Section 8.2, while the inclusion of a transportation module designated TRANS is considered in Section 8.3. The relative merits of alternative algorithms that could be adopted for solving the integrated models discussed in this book are examined in Section 8.4. Finally, some concluding remarks are then made in Section 8.5.

8.2 Inclusion of a National Econometric Module (NATLEC)

As mentioned above, a first potential extension to the integrated model developed in previous chapters would involve the inclusion of a national econometric module. This module would be responsible for generating national magnitudes for the components of final demand, consisting of household expenditures, investment, government expenditures, foreign exports and imports, and inventory change. In the initial run-through of the integrated model, these magnitudes help drive the interregional input-output, comparative cost-industrial complex, and programming modules. In addition, the NATLEC module would be responsible for projecting material prices, interest rates and other price variables which help drive the cost share equations within the factor demand (substitution) module and the investment equations within the regional econometric module.

Finally, the inclusion of this module would provide national aggregates against which checks can be made of the sum of regional estimates generated for similar variables -- eg. output, employment and investment by sector -- in other modules. In this way, the important problem of achieving top down/bottom-up consistency will be addressed explicitly. In this connection, it should be observed that the interconnections of the various non-policy modules with the NATLEC module can reflect various degrees of dominance of NATLEC. Where the national econometric module is of high quality, based on extensive data, and yields results highly consistent with theory one may permit the national magnitudes to serve as control totals not only for the initial round, but also for subsequent rounds. That is, by and large they would serve as binding magnitudes within which the various region-specific modules must operate. At the other extreme the NATLEC module may be poor in terms of both data base and quality of results consistent with theory. In this case the NATLEC module may serve only to provide the initial control total magnitudes -- and even then these magnitudes may be better derived from other partial research analyses. In this type of extreme case, the sum of the regional projections derived from operating the integrated model should be taken as valid for some magnitudes and hence used as the corresponding 'independent' variable values in other NATLEC equations.

In the in-between cases, some national magnitudes like investment in certain situations may be taken as the appropriate control totals -- especially when the data available for estimating regional investment equations are weak or where the results derived from the use of such equations are inconsistent with theory. In these cases, other national magnitudes, such as consumption expenditures and employment, may be estimated as sums of regional magnitudes derived in accord with a bottom-up approach. In these intermediate cases however, severe difficulties may arise since deriving national magnitudes as sums of regional magnitudes may seriously disturb the structure of the NATLEC module and in some cases destroy its consistency properties. In such intermediate cases then, an entirely new NATLEC module must be constructed which can be flexible enough to incorporate certain key national magnitudes derived using a bottom-up approach as the sum of their regional counterparts.

8.3 Inclusion of a Transportation Module (TRANS)

Another non-policy module which may be introduced pertains to transport network, commodity flow and transport cost analysis. For example we may use a model of the form discussed in Boyce and Hewings (1980). In this case, the TRANS module takes regional-sectoral outputs, regional-sectoral demands, energy shadow prices and unadjusted interregional sectoral flows from the comparative cost-industrial complex, input-output and programming modules. It then proceeds to minimize total system transportation cost $\sum_h \sum_J \sum_L {}^{JL}x_h \, {}^{JL}c_h$ (where ${}^{JL}x_h$ = amount of commodity h shipped from region J to region L, and ${}^{JL}c_h$ = transport cost per unit of h shipped from J to L), subject to traditional demand equals supply constraints stated in input-output form and an entropy constraint $C_h \leq \sum_J \sum_L {}^{JL}x \ln({}^{JL}c_h)$ for all h to allow for crosshauling of each commodity h.

Additionally, a network equilibrium submodel may be developed within the TRANS module to take into account congestion costs (a function of the flow on any link); and a modal choice submodel, as an extension of the route choice submodel.

Once composite costs over the transportation network are determined, the transportation costs are in a form satisfactory for use as inputs in the next iteration of the comparative cost-industrial complex and programming modules, and along with the TRANS estimates of flows, for use in revising trade coefficients in the interregional input-output module. Additionally, transport costs may be used as inputs in the gravity model component of the demographic module, and serve as major inputs into the factor demand (substitution) module to help derive a regionalised set of material and energy prices of greater validity.

8.4 Evaluation of Alternative Solution Algorithms

As mentioned previously, an important limitation of the model described in this book is that it represents a conceptual model only; no empirical studies have been completed to determine the parameter values associated with each component module. As a result, no formal proof can be provided that the iterative solution approaches recommended in Chapters 3-7 will lead to smaller and smaller discrepancies in the outcomes of the various modules from round-to-round. Nevertheless, other analysts have achieved favourable convergence results with models characterised by a size and complexity comparable to the present model. For example, see Almon, et al (1974), Ballard, et al (1980), Brown, et al (1972), Cherniavsky (1974), Courbis (1972, 1975, 1979, 1980, 1982a); Dixon (1979), Goettle, et al (1977), Higgs, et al (1981), Isserman (1980), Johnson, et al (1977), Olsen, et al (1977), and Treyz, et al (1980). Indeed some analysts have gone so far as to claim that with the improved capabilities of computer hardware and software over recent years, we have reached the situation where the technical problem of solving a large scale economic model should no longer be regarded a major limitation on their size and complexity (Drud, 1983).

A number of different solution algorithms have been employed to solve the above-mentioned models, however the majority fall into two classes -- Gauss-Seidel or Newton-type alogrithms.

Gauss-Seidel type algorithms assume that the model has been normalised, ie. that it has the form

$$x_i = f_i(x,y,P) \qquad i=1,...,n$$

where x_i is the ith endogenous variable, x is the whole vector of endogenous variables, y is the vector of exogenous variables and P is the vector of parameters. The model is solved by iterating on

$$x_i^{new} = f_i(x^{old}, y, P) \qquad i=1,\ldots,n$$

until x converges.

Newton-type algorithms assume the model has been written in the form

$$f(x,y,P) = 0$$

where f is a differentiable vector function with as many components as x. The model is solved by iterating on

$$x^{new} = x^{old} - \theta(\delta f/\delta x)^{-1} f(x^{old}, y, P)$$

where θ is a step-length.

A number of studies have been conducted recently providing comparative evaluations of the performance of these two classes of algorithms. For example, see Drud (1983), Norman, et al (1983), Manne (1985), Preckel (1985) and Don and Gallo (1987). The criteria employed in these evaluations include convenience (ie. how easy the algorithms are to use), flexibility (ie. how diverse a group of problems can be handled), efficiency (ie. what computer resources are used in converging on a solution), and robustness (ie. percentage of models successfully solved).

Newton-type algorithms were found to be more user independent since unlike Gauss-Seidel type algorithms they do not require normalisation and ordering of the equation's comprising the model. On the other hand, since they use derivatives, Newton-type algorithms often place higher demands on the modelling system. The matrix of derivatives must be stored and inverted, which can require large amounts of storage and computer time unless this matrix is relatively sparse. In terms of the efficiency criteria, the above-mentioned studies suggest large differences in efficiency and evidence of benefits from extensive numerical experiments and comparisons before selecting a final solution algorithm for any given model. Norman, et al (1983), for example, conducted experiments with a large scale model for which a Gauss-Seidel algorithm is found to solve the model at a significantly lower cost than a (modified) Newton algorithm. However, they applied Newton's method to the full set of variables in a simultaneous block and used sparse matrix techniques to solve Newton's equation. In contrast, Manne (1985) shows how the sparsity of the system can also be exploited to reduce the solution to a fixed point problem with a much smaller set of feedback variables. When this is done a (modified) Newton algorithm is found to solve the model at a significantly lower cost than a Gauss-Seidel algorithm. It is well known that the ordering of the equations can substantially affect the speed of convergence of the Gauss-Seidel algorithm; indeed some orderings may produce divergence for a given model. As a result, Don and Gallo (1987) explore the impact on solution algorithm efficiency of a number of different orderings of equations within a number of large scale economic models. Their findings once again suggest that a modified Newton-type algorithm is almost always superior to a Gauss-Seidel algorithm. In addition, the convergence properties of the Newton method were not affected by the particular orderings of the equations within the model. Finally, in terms of robustness Newton-type algorithms were generally found to be superior to their Gauss-Seidel equivalents.

A Newton-type algorithm will be employed when attempting to solve operational versions of the model described in this book. This algorithm has been developed as part of the GEMPACK software system and is fully documented in Codsi and Pearson (1988). This is the system currently employed for solving the ORANI model of the Australian economy (see Dixon, et al 1979) and a number of promising experiments have been conducted using this software for solving a skeletal version of the above-mentioned model.

8.5 Concluding Remarks

Although it is clear that the inclusion of both a NATLEC and a TRANS module would extend the range of issues which could be investigated using the integrated model, the demonstration of the operationality of such modules and/or the development of a sequence of steps by which they could be linked to the other non-policy modules awaits further research.

CHAPTER 9

POLICY ANALYSIS USING AN INTEGRATED MULTIREGIONAL MODEL

9.1 Introduction

The previous chapters have dealt with the various non-policy modules of an integrated multiregion model for the states of Australia.[1] However elsewhere (Isard and Smith, 1984, 1984a and 1984b; Smith, 1983 and 1986) we have suggested that the state of the art in multiregional modelling has reached the point where the policy variable can and must be incorporated more effectively. Such would enable us to make more accurate economic-demographic projections for each region of the system -- since it is inevitable that future policies (federal, state and local) will impact significantly upon the relative growth (and decline) of such regions. Additionally, and perhaps more importantly, it would enable us to examine more accurately the impacts of different proposed policy changes in a manner that recognizes responses likely to be forthcoming from other areas/levels of government and the private sector. This chapter begins to explore this new extension, and in particular addresses the question of how a conflict management - multipolicy formation module, designated INPOL, may be added to the model proposed in previous chapters.[2]

In Section 9.2 we point up several approaches that may be adopted when making policy mix projections in the Australian context. In Section 9.3 we discuss the need to link a given policy mix (set of policies) to the different non-policy modules of the integrated model. Finally, some concluding remarks are made in Section 9.4.

9.2 Illustrative Use of Policy Projection Framework

There are numerous ways in which outcomes of policy conflicts among the various interest groups pressing for different kinds of legislation can be projected. In this section we illustrate two sets of procedures for identifying (projecting) these outcomes. The first set, dealt with in section 9.2.1, rely heavily on an examination of the basic forces (pressures) influencing decision makers' policy choices. The second set, dealt with in Section 9.2.2, represents a more comprehensive and systematic approach involving an attempt to replicate the interplay of factors central to the policy formation process.

9.2.1 The Setting of Key Basic Forces: Historical Perspectives and Less Sophisticated Approaches

We first focus on the basic forces (pressures) influencing the choice of a policy mix by a single decision-making (governmental) unit, and suggest the use of the following procedure.

a. Single Decision-making (Governmental) Unit

Step 1: *Estimate change in pressures for different types of polices from year 1988 to the year of projection, say 2000*

As a start, crude trend analysis might be employed. It would then be necessary to find relevant indicators of past trends in pressures for change in each of the policy areas identified as of key importance to the nation over the period 1988-2000. We make a first attempt at doing this in Table 9.1. Here column 1 contains a listing of key policy issues, while column 2 contains a listing of potential indicators of past trends in pressures (for change) in each of these policy areas. However, as is well known, projections on the basis of past trends alone can often be improved upon. Therefore we list in column 3 potential indicators of changes in past trends. In some instances the latter suggest a dampening of past trends, and in others an upward adjustment.

Step 2: *Identify the most likely position on each policy in the year 2000, ignoring interdependence of policies*

It is one thing to develop a set of indicators showing the direction and level of pressures for change in a given policy position. It is another

Table 9.1 Key Policy Pressure Indicators

Key Policy Issue Indicator	Potential Indicators of Past Trends in Pressures for Policy Change	Potential Indicators of Changes in Past Trends
I. Pressure for Tariff Reduction	(1) number of times legislation attempting to reduce tariff levels has been introduced into federal parliament (2) level of support/opposition to the legislation in (1), as indicated by the number of representatives voting for it (or speaking for it, should a vote not be taken) . . .	(a) proportion of total employment in agricultural, services, mining and other non-manufacturing sectors which are anticipated for year 2000 (for each region and the nation) -- these sectors being less dependent on tariff protection for their economic survival (b) proportion of representatives who come from regions (states) which are anticipated not to be heavily dependent on protected manufacturing industries by year 2000, and hence whose constituents are more likely to support tariff reduction (c) need for tariff protection, as indicated by changes in the 'competitive' position of domestic producers which are anticipated by year 2000 . . .
II. Pressure for Increased Regulation of Mineral Resource Development Projects	(1) levels of urbanisation, formal education completed per capita and income per capita (in each region and the nation) (2) number of supporters of groups fighting for conservation of wilderness areas (in each region and the nation) (3) number of contributors and increase in total contributions to environmental protection groups (in each region and the nation)	(a) best estimate of geologists and mining engineers of the quality and quantity of diverse mineral resources in both sensitive (wilderness, aboriginal, etc) and non-sensitive areas (b) need for foreign exchange earnings, as indicated by the size of balance of payments deficits anticipated by year 2000 (c) changes in the extent of reliance on mineral exports as a foreign exchange earner, as anticipated for year 2000 .

Table 9.1 (continued)

		(4) community recognition of aboriginal rights and the need to protect them, as indicated by results of opinion polls and types of legislation being passed in federal and state parliaments	
III.	Pressure for Imposition of Increased Taxation on Mining Industry	(1) number of times legislation attempting to increase the level and extent of royalties levied by state governments has been introduced into each state parliament	(a) need for foreign exchange earnings, as indicated by size of balance of payments deficits anticipated by year 2000
		(2) level of support/opposition to the legislation in (1)	(b) changes in the extent of reliance on mineral exports as a foreign exchange earner, as indicated by the proportion of total export earnings accounted for by mining and/or mineral-based products in year 2000
		(3) number of times legislation attempting to increase the level and extent of export levies imposed by the federal government has been introduced into parliament	(c) best estimates of geologists and mining engineers of the quality and quantity of diverse mineral resources in remote areas and hence requiring extensive infrastructure provision by state governments
		(4) level of support/opposition to the legislation in (1)	(d) need for additional tax revenue, as indicated by size of federal and/or state government deficits anticipated in year 2000
		(5) level of foreign ownership and control of mining projects	

Table 9.1 (continued)

IV.	Pressure for Increased Levels of Foreign Immigration	(1) number of times legislation attempting to relax immigration regulations have been introduced into federal parliament	(a) shortages of labour anticipated to arise by year 2000 -- either in total or by skill category
		(2) level of support/opposition to the legislation in (1)	(b) proportion of total employment in industries which could exploit cheap skilled/unskilled labour -- as anticipated by year 2000
		(3) number of times legislation attempting to increase government support to migrants (assisted passage, hostel provision, english language education, welfare payments upon arrival) has been introduced into federal parliament	(c) proportion of representatives who come from regions (states) which are anticipated to be heavily dependent on industries in (b)
		(4) level of support/opposition to the legislation in (3)	
		(5) strength of local trade unions (measured in terms of members)	
		(6) relative importance of industry that could exploit cheap skilled or unskilled labour	
V.	Pressure for Reduced Foreign Ownership and Control of Australian Enterprises	(1) number of times legislation attempting to reduce foreign ownership and control has been introduced into federal parliament	(a) need for foreign exchange earnings, as indicated by the size of balance of payments deficits anticipated by year 2000
		(2) level of support/opposition to the legislation in (1)	(b) need for capital inflow to offset balance of payments (current account) deficits anticipated by year 2000
		(3) proportion of local savings being channeled into risky ventures (eg. mineral development)	(c) coefficient of industrial diversification anticipated in year 2000 (the greater the diversification, the less the need to rely on foreign capital to introduce new types of economic activity)
			(d) national propensity to save anticipated in year 2000

Table 9.1 (continued)

VI. Pressure for Change in the Level of Federal Expenditures and its Allocation among Programmes	(1) defence expenditures per capita (2) social welfare expenditures per capita (3) education, health and other social infrastructure expenditures per capita	(a) per capita income levels anticipated in year 2000 (b) age composition of population anticipated in year 2000 (c) female workforce participation rate anticipated in year 2000 (d) best estimates of extent of dependence on US and UK as sources of defence and security in year 2000

thing to project how decision makers will respond to these pressures; yet such a projection is required in order to derive a point estimate of the most likely position on each policy. On the one hand, the analyst may make his best guess of each of these positions. Alternatively, the analyst may conduct a Delphi-type policy workshop with a group of experts and seek (after a number of rounds of interaction with them) consensus on a most likely position on each policy, given the set of pressures and changes in them identified in Step 1.[3] Finally, a Saaty type approach may be adopted wherein a group of experts is assembled for each policy area. Each of these experts is asked to make pairwise comparisons concerning the relative likelihood of a finite set of positions which may be adopted on his particular policy issue by the given decision-making (governmental) unit. His implied estimate of the most likely of these positions can then be identified using the eigenvector technique.[4] The most likely position estimates on any given policy area can then be averaged (either directly or indirectly) over all experts on that policy area.[5] This yields a first very crude "most likely" set of policies (policy mix).

Step 3: *Identify the policy mix most likely to be adopted in year 2000 taking into account policy interdependence*

As is well appreciated, policies are interdependent. Hence an attempt to identify a most likely position in each policy area independent of others would in general be expected to generate inconsistencies. For example, a pair of most likely policies involving lower tariffs and no change in mineral export levies by the federal government may be inconsistent with policies of reduced federal taxes and increased levels of federal government programs. And so on. These inconsistencies may then be pointed out to the above mentioned experts on each policy area, and a revised set of most likely policy positions generated. Alternatively another group of experts more knowledgeable on the give and take in political processes and the development of legislation may be consulted. The latter experts may be able to suggest reasonable compromises among the inconsistent (and thus necessarily conflicting) policies identified in Step 2 -- and thereby identify a most likely set of policies of greater consistency.

b. Many Interest Groups Influencing a Single Decision-making (Governmental) Unit

Another approach which is more extensive, but more reasonable, focuses first on the basic forces (pressures) acting on each of the major interest groups influencing the given decision-making unit. It comprises the following steps:

Step 1:

With assistance from experts on social, political and economic forces identify the set of most likely major interest groups (including major sectoral lobby groups) in operation at the given level of government over the period 1988-2000, and in year 2000.

Step 2:

Choose a team of experts on the perspectives, goals, beliefs and behavior patterns of each of these relevant interest groups.

Step 3:

Present each team identified in Step 2 with information on:

(a) the existence, perspectives, goals, etc. of other interest groups (both those who are likely to take opposite positions on any given policy issue, and those likely to take complementary positions); and

(b) the set of pressures for change in each policy area, as projected for the period 1988-2000 and in year 2000 using the approach discussed in Step 1 of section a above.

Step 4:

Ask each individual expert in any given team identified in Step 2 to state his estimate of:

(a) the most likely formal position on each policy issue, and

(b) the most likely optimal mix of policies for the interest group on which he is expert.

For each interest group the analyst can then generate a (rough) average of the opinions of the individual experts on that group.

Step 5:

Bring together another set of experts, namely those knowledgeable on the give-and-take of political processes and the development of legislation. Present them with (1) the average estimate of the most likely formal policy position of each interest group, and (2) the average estimate of each group's most likely optimal mix of policies.

Step 6:

Hold a brainstorming or workshop-type interaction session[6] with the experts identified in Step 5. As a result of this session, identify a finite set of reasonable policy mixes over which debate and discussion among political leaders (i.e. legislators) may range in year 2000.

Step 7:

Identify the most likely policy mix for year 2000 by either:

(a) asking the experts identified in Step 5 to make pairwise comparisons of the relative likelihood of adoption of the policy mixes identified in Step 6. (That is, ask them how many "times" more likely is it that policy mix a_j will be adopted than policy mix $a_{j'}$ -- or perhaps more realistically ask them to use the Saaty scale ranging from one to nine when making these comparisons). The most likely policy mix can then be derived from each individual expert's pairwise comparisons matrix via use of the eigenvector technique.[7] These individual estimates can then be averaged to obtain the overall most likely policy mix; or

(b) asking experts identified in Step 5 to employ a crude split-the-difference (or simple rank-oriented) procedure over the most preferred policy mixes within the restricted set of policy mixes identified in Step 6.

The latter approach may be useful should political leaders in year 2000 be expected to be relatively unsophisticated; however it may be that the two approaches lead to similar outcomes.

A number of alternative procedures can be developed as variants of the approach described in this section. For example, one variant would allow feedback from operating the integrated multiregional model to influence the experts' estimates (derived in Step 4) of the most likely formal position and optimal policy mix for that interest group on which they are knowledgable. Another procedure would allow feedback from running the integrated model for each of the reasonable compromise policy mixes identified in Step 6. The results are then given to experts in order to assist them in making their pairwise comparisons in Step 7.

9.2.2 A More Comprehensive and Systematic Approach Toward Policy Projection

Although the projection procedures in the above section may be satisfactory or the best an analyst can do given limited resources and time, it is possible to outline a more comprehensive and systematic approach toward policy projection which is more sensitive to the cultural structure and practices of a given system (nation) under study. It also may be able to reflect more adequately the interplay of factors that in reality may determine the future policy mix.

a. Basic Elements of a Policy Choice Using the Saaty Approach : Single Decision-making (Governmental) Unit

We now specify the basic elements in a policy choice by a given government unit (say the Federal government). They are:

(1) interest groups influencing the decisions of the given government unit z_i (i=1,...,I) -- see oval 4 of Figure 9.1,

(2) objectives of importance to one or more interest groups v_h (h=1,...,H) -- see oval 3 of Figure 9.1,

(3) outcome dimensions (as derived from the operation of the integrated model) o_g (g=1,...,G) -- see oval 2 of Figure 9.1,

(4) policy proposals (joint actions) a_j (j=1,...,J) -- see bold ovals A and E of Figure 9.1,

(5) policy instruments (specified in a form amenable for use in the integrated model) μ_f (f=1,...,F) -- see oval 1 of Figure 9.1.

Given these elements, there exist many ways by which to evaluate the different policy mixes that may be proposed for the year 2000 in terms of their outcome implications as determined via the operation of the non-policy modules of an integrated multiregion model. With reference to Figure 9.1, one possible set of steps may be as follows.

Step 1: *Selection of an Initial Set of Policy Mixes to be Considered*

While many policy proposals (mixes or joint actions) may be considered, in practice only a few can be examined for outcome implications with the use of the non-policy modules of the integrated multiregion model. The activity involving the selection of these few proposals (which may be viewed as <u>setting the agenda</u>) via diverse compromises is depicted by the bold oval A where the political leaders, representatives of diverse interest groups and other behaving units interact.[8]

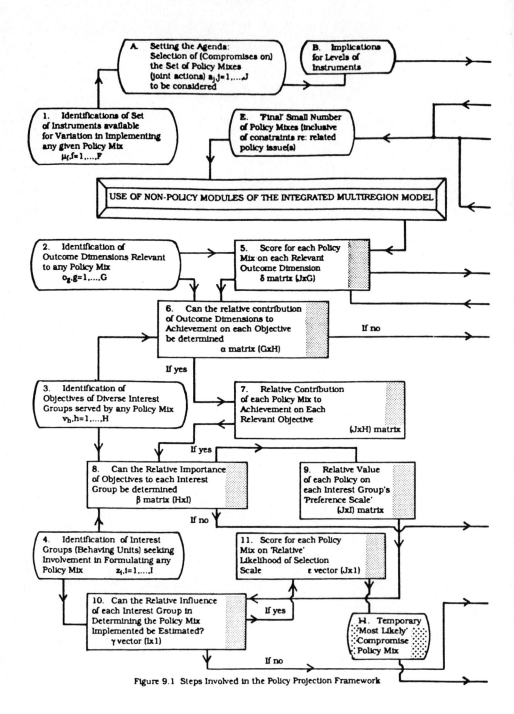

Figure 9.1 Steps Involved in the Policy Projection Framework

POLICY FORMATION MODULE

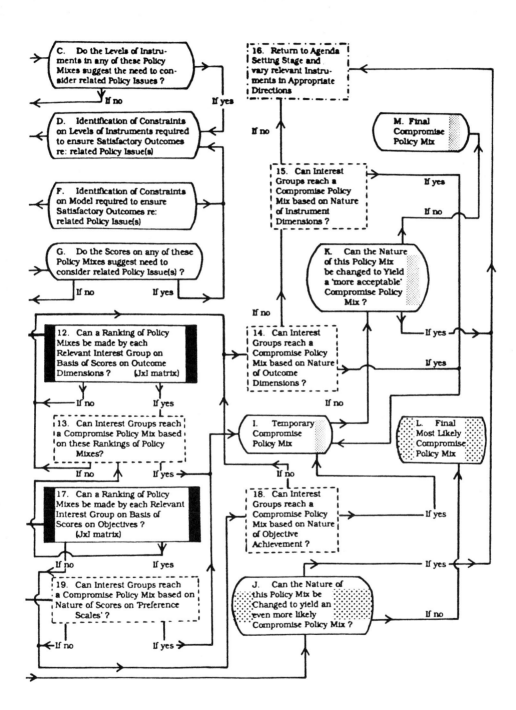

Step 2: *Identification of Nature and Level of Instruments Required to Implement the Given Policy Mixes*

In bold ovals B, C, D and E respectively we:

(1) determine the set of instruments μ_f (f=1,...,F) and levels of each required to effect each given policy mix (proposal) a_j (j=1,...,J):[9]

(2) consider whether the levels of instruments identified in (1) suggest the need to consider related policy issues not covered by these policy mixes;[10]

(3) if there is such a need, determine (identify) the constraints on the levels of the instruments required to ensure satisfactory outcomes regarding related policy issues; and

(4) eliminate those policy mixes which cannot meet the constraints identified in (3). This yields a JxF matrix recording level of the various instruments μ_f (f=1,...,F) required to implement the final small number of policy mixes a_j (j=1,...,J) to be considered in subsequent steps.

Step 3: *Identification of Outcome Implications of Each Policy Mix*

Each policy mix a_j can now be run through the non-policy modules of the integrated multiregion model indicated by the long rectangular box in Figure 9.1. This step is achieved by setting each relevant policy instrument at the corresponding level identified in Step 2. This yields a score for each policy mix (proposal) a_j (j=1,...,J) on a number of relevant outcome dimensioins o_g (g=1,...,G).[11] See the shaded box 5 in Figure 9.1. We can then construct an outcome matrix (designated δ) of order JxG with typical element o_j recording the score of policy mix j on outcome dimension g. That is:[12]

$$
\begin{array}{c}
\begin{array}{cccc} 1_o & 2_o \ldots & g_o \ldots & G_o \end{array} \\
\begin{array}{c} a_1 \Rightarrow o_1 \\ a_2 \Rightarrow o_2 \\ \cdot \\ \cdot \\ \cdot \\ a_j \Rightarrow o_j \\ \cdot \\ \cdot \\ \cdot \\ a_J \Rightarrow o_J \end{array}
\left[
\begin{array}{cccc}
1_{o_1} & 2_{o_1} \ldots & g_{o_1} \ldots & G_{o_1} \\
1_{o_2} & 2_{o_2} \ldots & g_{o_2} \ldots & G_{o_2} \\
\cdot \cdot \cdot & \cdot & \cdot \cdot \cdot & \cdot \\
\cdot \cdot \cdot & \cdot & \cdot \cdot \cdot & \cdot \\
\cdot \cdot \cdot & \cdot & \cdot \cdot \cdot & \cdot \\
1_{o_j} & 2_{o_j} \ldots & g_{o_j} \ldots & G_{o_j} \\
\cdot \cdot \cdot & \cdot & \cdot \cdot \cdot & \cdot \\
\cdot \cdot \cdot & \cdot & \cdot \cdot \cdot & \cdot \\
\cdot \cdot \cdot & \cdot & \cdot \cdot \cdot & \cdot \\
1_{o_J} & 2_{o_J} \ldots & g_{o_J} \ldots & G_{o_J}
\end{array}
\right] \\
\text{JxG}
\end{array}
$$

At this point we need to determine whether or not the outcome scores for any of these policy mixes suggest the need to consider implications for related policy issues not included in that policy mix.[13] See bold oval G. If not, we can proceed to Step 4. Otherwise, we may need to change the levels of some of the instruments (bold oval D) associated with the given policy mixes. We may even need to impose other constraints[14] (bold oval F) which could result in the elimination of one or more policy mixes from further consideration. The integrated model must then be rerun for each non-eliminated policy mix under the new constraints. This then yields a new δ matrix.

Step 4: *Identification of the Relative Contribution of Each Outcome Dimension to Achievement on Each Objective*[15]

This can be done via the use of the Saaty procedure for determining group priorities. See shaded box 6. For example, for a given objective v_h we might (after consultation with relevant experts) obtain a pairwise comparisons matrix (GxG) of the form:[16]

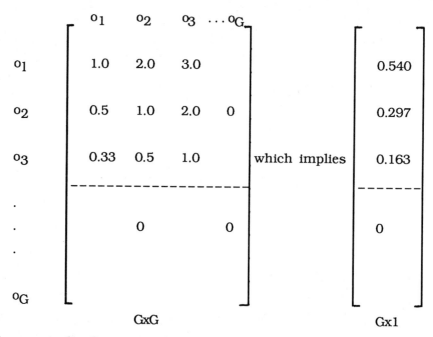

when normalised.

As indicated above, this yields a Gx1 vector of "weights" showing the relative contribution of outcome dimension $o_g(g=1,...,G)$ to the achievement of objective v_h.[17] For each of the H objectives a comparable Gx1 vector can be derived, and when these results are recorded in a single table this yields a matrix (designated α) of order GxH.

Step 5: *Identification of the Relative Contribution of Each Policy Mix to Achievement on Each Objective*

This can be done by premultiplying the α matrix of order GxH derived in Step 4 (shaded box 6) by the δ matrix of order JxG derived in Step 3 (shaded box 5) to obtain, as noted in shaded box 7, a JxH matrix. The typical element of this latter matrix gives the relative contribution of policy mix j (j=1,...,J) to the achievement of objective h(h=1,...,H).

Step 6: *Identification of the Relative Importance of Each Objective to Each Interest Group*[18]

This can be done via the use of the Saaty procedure for determining group priorities. See shaded box 8. For example, for a given interest group z_1 we might (after consultation with relevant experts) obtain a pairwise comparisons matrix (HxH) of the form:[19]

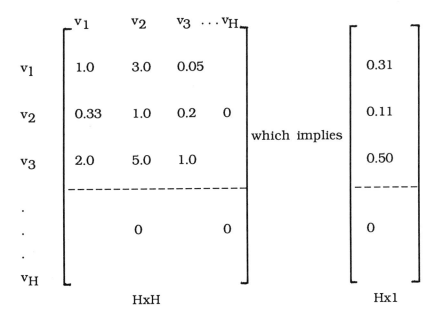

when normalised.

As indicated above, this yields a Hx1 vector of "weights" showing the relative importance of objectives $v_h (h=1,...,H)$ to the interest group z_i.[20] For each of the I interest groups a comparable Hx1 vector can be derived, and when these results are recorded in a single table this yields a matrix (designated ß) of order HxI.

Step 7: *Identificaton of the Relative Value of Each Policy Mix for Each Interest Group*

We first premultiply the ß matrix of order HxI derived in Step 6 (shaded box 8) by the α matrix of order GxH derived in Step 4 (shaded box 6) to obtain a GxI matrix which indicates the relative importance of each outcome dimension to each interest group. We can then premultiply this GxI matrix by the δ matrix of order JxG derived in Step 3 (shaded box 5) to obtain, as noted in shaded box 9, a JxI matrix. The typical element of this latter matrix gives the relative value of polixy mix $j(j=1,...,J)$ to each interest group $i(i=1,...,I)$.

Step 8: *Identification of the Relative Importance (Influence) of each Interest Group in the Policy Formation Process*[21]

This can be done via the use of the Saaty procedure for determining group priorities. See shaded box 10. For example, after consulting with experts on the policy formation process we might obtain a pairwise comparisons matrix of the form:[22]

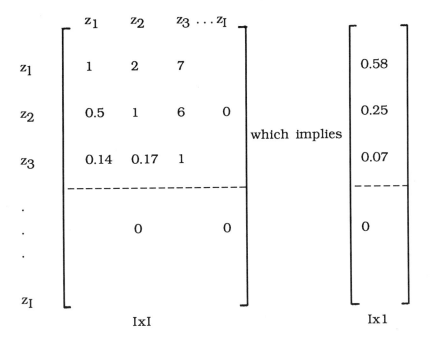

when normalised.

As indicated above, this yields a Ix1 vector of "weights" showing the relative importance (influence) of each interest group $z_i(i=1,...,I)$ in selecting the policy mix to be put into effect by the given level of government. We designate this the γ vector.

Step 9: *Identification of the Relative Likelihood of Adoption of Each Policy Mix*

We next multiply the JxI matrix showing the relative value of each policy mix to interest groups derived in Step 7 (shaded box 9) by the γ vector of order IxI derived in the previous step (shaded box 10). As noted in the shaded box 11, this yields a Jx1 vector (designated ε) whose typical element gives the relative likelihood of selection (importance) of policy mix $a_j(j=1,...,J)$ by the given level of government.[23] This is the desired outcome, and it may appear at first glance that we can now conclude that the policy mix (proposal) with the highest relative value be designated the 'most likely' compromise policy proposal. See the oval H. However, we regard this conclusion as only temporary and suggest the adoption of one more step.

Step 10: *Resetting of the Agenda in the Light of the Results of Previous Steps*

Next we may enquire (or the policital leaders may consider) whether the nature of the most likely policy mix (identified in Step 9) can be changed to yield an even more likely "compromise" policy mix. See the oval J. If not we (they) accept the results in shaded box 11 as final -- thus moving to oval L. If yes, we (they) in effect choose to return, via box 16 to the agenda setting stage (ovals A through E), introduce an adjusted policy mix and go through the entire process again. However, if the Saaty procedure is once again employed some new problems and inconsistencies arise. We cannot discuss these further here due to limited space.

b. Extension to Many Decision-making (Governmental) Units: Modifications to the Basic Elements of a Policy Choice

The previous section treated the basic elements of a policy choice by a single governmental unit, yet frequently more than one such unit is involved in the kind of policy analysis that regional scientists need to conduct. In what follows, we have in mind the choice of a national policy mix wherein many state (provincial, prefectual) and local (county, etc.) government units have influences through governors (premiers), mayors, and the like. However, the framework can easily be extended to consider interdependent policy formation at the state and local levels themselves -- as well as at all levels simultaneously. The following modifications are then required to the general framework outlined in part a of section 9.2.1.

(1) We must now be explicit about the levels of government $\emptyset_d(d=1,...,D)$ involved in formulating the given policy mix. This in turn requires that within the set of relevant interest groups we allow for the possibility of a hierarchical structure $z_{id}(i=1,...,I; d=1,...,D)$.[24]

(2) We must now conduct Steps 6 and 7 for each relevant group z_{id}. As a result, the matrices in the shaded boxes numbered 8 and 9 of Figure 9.1 change to order (HxID) and (JxID) respectively.

(3) The remaining steps (8-11) are changed as follows.

Step 8': *Identification of the Relative Importance (Influence) of each Interest Group on the Policy Position adopted by each Government Unit*

This can be done via use of the Saaty procedure for determining group priorities. For example, for a given government unit $\emptyset_d$ we might (after consultation with relevant experts) obtain a pairwise comparison matrix of order IDxID. Using the eigenvector technique this yields a IDx1 vector of weights showing the relative importance (influence) of each interest group $z_{id}(id=1,...,ID)$ to this government unit. For each of the D levels of government, a comparable IDx1 vector can be derived, and when these results are recorded in a single table this yields a matrix (designated η) of order IDxD.

Step 9': *Identification of the Relative Value of Each Policy to Each Government Unit*

We first multiply the (JxID) matrix showing the relative value of each policy mix to each interest group derived in the revised Step 6 by the η matrix of order (IDxD) derived in the previous step. This yields a JxD matrix whose typical element gives the relative value attached to policy mix $j(j=1,...,J)$ by each government unit $d(d=1,...,D)$.

Step 10': *Identification of the Relative Importance (Influence) of each Government Unit in the Policy Formation Process*

This can be done via the use of the Saaty procedure for determining group priorities.[25] For example, after consulting with experts on the policy formation process we might obtain a pairwise comparisons matrix of order DxD. Using the eigenvector technique this yields a Dx1 vector of "weights" showing the relative importance (influence) of each government unit $\phi_d(d=1,...,D)$ in selecting the policy mix to be put into effect. We designate this the θ vector.

Step 11': *Identification of the Relative Likelihood of Adoption of each Policy Mix*

We next multiply the JxD matrix (derived in Step 9') showing the relative value of each policy mix to government units by the θ vector derived in the previous step. This yields a Jx1 vector designated ε' whose typical element gives the relative likelihood of selection (implementation) of policy mix $a_j(j=1,...,J)$ by the given government unit.[26]

9.2.3 Difficulties and Qualifications in Policy Projection Framework

The framework for projection outlined above obviously simplifies reality. One major point at which it does is in its implicit assumption of symmetry with regard to the answers of the questions posed in Figure 9.1. However, we must recognize that asymmetries can be expected to arise in the answers projected for one or more of these questions. Hence we make some initial suggestions of how this framework can be modified to accommodate these asymmetries. The latter arise:

(a) in the extent to which the Saaty procedure can be used to yield relative weights (shaded boxes 6, 8 and 10),[27]

(b) in the extent to which the analyst is able to project utility values on comparable scales for each interest group/level of government (boxes with bolded sides numbered 12 and 17, and their substitutes),[28] and

(c) in the extent to which the analyst considers that interest groups will be able to focus on the utility space and compromises there-in (dashed boxes 13 and 19), on the objective achievement space and compromises therein (dashed box 18), on the outcome dimension space and compromises therein (dashed box 14), or (at the minimum) on instruments and compromises therein (dashed box 15).[29]

The second major point at which the above framework simplifies reality is that it abstracts from the formation, disruption, reformation and other dynamic aspects of coalition structure development. Interest groups are, in essence, treated as independent and stable entities. While major advances have been (and continue to be) made in game theory by mathematicians, operations researchers, mathematical economists and the like, still not enough practical knowledge has been accumulated to allow for the meaningful projection of coalition structures over the medium to long-run. Clearly, there is an urgent need for more research in this area.

9.3 Relationship of INPOL Module to other Modules Comprising the Integrated Model

The question which remains to be addressed explicitly is how exactly does the conflict management and multipolicy formation module feed into and respond to outputs from the other modules comprising the integrated model. The nature of these linkages have been illustrated in a general way via the sets of arrows leading into and out of the long, narrow, rectangular box designated "Use of Non-Policy Modules of the Integrated Multiregion Model" in the center of Figure 9.1. The specific ways and exact points at which these linkages will occur depend on the specific policy (or set of policies) under consideration. For a detailed outline of such linkages for (a) a reduction in the level of tariff protection, (b) a change in mineral development policy, (c) an increased foreign immigration intake, and (d) a change in federal government fiscal priorities, see Smith (1983:484-526).

9.4 Concluding Remarks

In this chapter, we have completed our presentation of the integrated multiregion model that provides the focus of this book.

Moreover, we have explicitly recognized that no projection can be made for a future year (say 2000) without a point projection of the policy mix that is most likely to be adopted in that year. Hence we have illustrated ways in which the analyst might arrive at a most likely policy mix to be used. Here, more than at any other place within the integrated model, the subjective judgements of experts must be relied upon. The extent to which the analyst chooses to rely on such judgement is his choice, but he <u>must</u> introduce an estimated most likely policy mix -- and in doing so he should be explicit about the assumptions underlying the selection of this particular mix rather than a multitude of others. Any hiding behind implicit assumptions on this important aspect of the integrated model's operation can be argued to reflect the limited knowledge or ability of the analyst -- or scholarly dishonesty!

NOTES

[1] That is these modules are concerned with projecting economic and/or demographic magnitudes, rather than the mix of policies adopted in a given year.

[2] It should be emphasised that we have only just begun to develop comprehensively and systematically the analysis required to make this module comparable to the others in terms of rigour of supporting theory, depth of operational analysis, and quality of empirically testable hypotheses. (See, for example, Isard and Smith (1982)). Nonetheless, we can begin to indicate the kinds of approaches which may be usefully employed within this module.

[3] See Isard and Smith (1982, Appendix 10A) for a discussion of the Delphi method.

[4] See Isard and Smith (1982, Chapter 5) for a discussion of the Saaty approach.

[5] One can pursue the Saaty procedure one step further and derive relative weights to apply to these experts' estimates. (This would require, however, that some knowledgable person make pairwise comparisons of the different experts' levels of expertise).

[6] See Isard and Smith (1982, Chapter 10) for a discussion of these forms of interaction.

[7] While the Saaty method of determining group priorities is among the better approaches for projecting the relative importance of policies and a "most likely" policy mix, it is not without serious shortcomings. For example the interactive non-linear relations of policy outcomes, policy objectives, interest groups, etc., cannot be handled by it no matter how many levels are included in its hierarchy. Also it does require that each expert's pairwise comparisons matrix contain a fairly high degree of consistency before it can be meaningfully employed. Further, the individual experts' estimates should not differ by unreasonable amounts. For further discussion, see Isard and Smith (1982, Chapter 5).

[8] For further relevant discussion, see Smith(1983,Chapter 10).

[9] For example, the policy mix may relate to military expenditures, energy resource development, low income housing provision, transportation network extensions, etc. In this situation the relevant set of instruments would include military procurement by region, employment on military bases by region, subsidies on oil shale installations by region, dollar expenditures on federal housing by region, and expenditure on mass transit by region.

[10] For example, should the level of tariffs be reduced across-the-board by 25 percent in order to effect a trade policy, then we may need to consider the introduction of a manpower planning policy to avert the undesirable employment implications of this tariff change.

[11] Typical outcome dimensions would be regional employment and/or income levels, unemployment rates, air pollution levels, etc.

[12] Note that each row represents the outcome vector corresponding to a given policy mix, and that each column has been normalised (ie. converted to a common scale).

[13] For example, examining the results recorded in the δ matrix may reveal that a given policy mix (say a_j·) leads to an undesirably high level of unemployment in a given region (or for a particular class of workers). This may then suggest the need for the simultaneous adoption of a region (or class) specific employment development program.

[14] For example, such constraints may relate to outcome dimensions such as oil imports, capital available for investment in housing and infrastructure, and disparities in regional per capital income. In some situations, political leaders (or their right hand men) may be unable to agree on a particular set of policy mixes to be considered, yet able to specify a minimum acceptable target score on each outcome dimension. The analyst may then be able to identify a set of policy mixes which are each capable of meeting these target scores.

[15] If such relative contributions cannot be identified using the Saaty procedure, then we must consider other alternatives. For example, (as indicated in boxes 12 and 13, respectively) should a ranking of policy mixes (or proposals) be available (or able to be constructed for each interest group on the basis of the outcome scores identified in step 3, then the analyst may consider the use of rank-oriented conflict management procedures, which yield a temporary compromise policy proposal, as indicated in the light dotted oval I). Where such rankings are unavailable, the analyst may focus on the outcome scores or instrument dimensions directly and (via the use of one or more reasonable conflict management procedures) reach agreement on a set of outcomes/instrument levels. These options are indicated in boxes 14 and 15, respectively. If successful, they too yield a temporary compromise policy proposal (light dotted oval I). On the other hand, should none of these approaches work, then the analyst will need to return via box 16 to the agenda setting stage (bold oval A) and consider a revised set of policy mixes. For further details, see Smith (1983, 547-558).

[16] Note that the 3.0 in the cell o_{13} implies that o_1 makes a "weakly more important" contribution to the given objective than o_3, and in particular that it makes a three times more important one than o_3, when using Saaty's scale and method. For a full discussion of the scale used in

making such pairwise comparisons, see Saaty and Khouja (1976) and Isard and Smith (1982:148-50).

[17] For example, the 0.540 and 0.297 in the first and second rows of this vector indicate that, compared with a unit of o_2 a unit of o_1 makes a (0.540/0.297) times greater contribution to objective v_h.

[18] If such relative importances cannot be identified using the Saaty procedure, then we must consider other alternatives. For example, (as indicated in box 17) should a ranking of policy mixes (or proposals) be available (or able to be constructed) for each of the interest groups on the basis of the objective achievement scores identified in Step 5, then the analyst may consider the use of rank-oriented conflict management procedures. (This yields a temporary compromise policy proposal, as indicated in the light dotted oval I). Where such rankings are unavailable, then the analyst may focus directly on the objective achievement scores and (via the use of one or more reasonable conflict management procedures) reach a compromise set of objective achievements. This option is indicated in box 18 and, if successful, it too yields a temporary compromise policy proposal. On the other hand, should neither of the above approaches work, then the analyst will need to consider one or more of the options discussed in footnote 15 above. For further details, see Smith (1983, 558-563).

[19] Note that the 5.0 in the cell v_{23} implies that interest group j considers that v_2 is of "strong" importance relative to v_3, and in particular is five times as important as v_3 when using Saaty's scale and method.

[20] For example, the 0.31 and 0.11 in the first and second rows of this vector indicate that for interest group z_i a unit of objective v_1 is regarded as (0.31/0.11) times more important than a unit of objective v_2.

[21] If such relative influence (importance) cannot be identified using the Saaty procedure, then we must consider other alternatives. For example, (as indicated in box 19 outlined by dashes), the analyst may consider the use of outcome-oriented conflict management procedures based on the relative preference values derived for each interest group in step 7. These procedures yield a temporary compromise policy proposal (light dotted oval I). If no acceptable outcome-oriented CMP's can be found, then one or more of the approaches discussed in Footnotes 15 and 18 above need to be employed. For further details, see Smith (1983, 543-564).

[22] Note that the 7.0 in the cell z_{13} implies that, for the given government unit, interest group z_1 is considered to have "demonstrated" influence relative to z_3 -- and in particular is seven times as important as z_3 when using Saaty's scale and method.

23Mathematically, we may summarize the operations performed up to this point as

$$[\varepsilon] = [\delta] \quad [\alpha] \quad [\beta] \quad [\gamma]$$
$$Jx1 \quad JxG \quad GxH \quad HxI \quad Ix1$$

24In this hierarchy, an environmental protection (E.P.) group operating at the national (federal government) level would be considered different from an E.P. group operating at the state or local level. This is so since the latter may place a greater weight on environmental quality in its particular state or locality than its national counterpart and thus place a different set of weights on the total set of objectives.

25If such relative influence (importance) cannot be identified using the Saaty proceudre than we must once again consider alternatives. See Smith (1983, 567-568) for further details.

26Mathematically, we may summarize this modified set of operations as:

$$[\varepsilon'] = [\delta] \quad [\alpha] \quad [\beta'] \quad [\eta] \quad [\theta]$$
$$Jx1 \quad JxG \quad GxH \quad HxID \quad IDxD \quad DxI$$

27For example, after consultation with relevant experts the analyst may only be able to determine the relative importance of objectives to some but not all interest groups. And so on. In this situation we suggest:

(1) that the analyst treat the answer to the question in the corresponding shaded box as if no, however

(2) that he follow through the derivation of relative weights for all those outcome dimensions, objectives, interst groups and/or government units for which such is possible. This then provides additional information which, in general, can be expected to prove useful to the analyst when he projects the interest groups' rankings of policy mixes (boxes with shaded sides numbered 12 and 17) and compromises therein.

28For example, the analyst may be able to project cardinal (utility) valuations for one interest group, yet only ordinal (utility) valuations (rankings) for the other interest groups. In this situation, we suggest that the analyst consider compromise procedures which require the least amount of information.

27For example, some but not all interst groups may be considered able to reach a compromise on the basis of objective achievements. In this situation, we suggest:

(1) that the analyst treat the answer to the question in the corresponding dashed box as if no, however

(2) that in proceeding to the next box he should not ignore information available to him in the former box, since it may prove useful in pointing up those procedures (or solutions to procedures) which can be expected to be rejected by the interest groups who could focus on the dimension requiring the larger amount of information.

CHAPTER 10
CONCLUSIONS

This study has addressed the issues involved in designing a system of multiregional models which could be used for policy analysis in the Australian context. A good deal of attention has been focussed on the development of the various component modules -- interregional input-output, comparative cost-industrial complex, interregional programming, demographic, regional econometric, factor demand (substitution) and conflict management-multipolicy formation. However, equally important has been: (1) the specification of a sequence of steps by which the non-policy modules could be operated, and (2) the identification of linkages to and from the conflict management-multipolicy formation module.

In Chapter 2, we dealt with the core component of the integrated model -- namely that which determines (projects) the level of economic activity in each region. The economic base approach was considered first, but in general it was dismissed as inferior should the resources exist to support the operation of an input-output module. A number of variants of the input-output approach were outlined, and the conclusion drawn that an interregional input-output table was the most appropriate for use in an integrated model.

In Chapter 3, we dealt with the development of a comparative cost-industrial complex module which would be capable of forming a component of an integrated multiregion model. An outline is given of the steps involved in conducting both comparative cost and industrial complex analyses for a system of regions. We then discussed how the results of these analyses can be used to provide projections of the regional production/trade patterns in cost-sensitive industries while retaining the use of an interregional input-output model for projecting production/trade patterns for other sectors of the multiregional economy. Although the operationality of this module has yet to be proven in the Australian context, we feel confident that such can and must be done in order to project overseas exports of cost-sensitive minerals and mineral-based products.

In Chapter 4, we dealt with the development of a linear programming module which would be capable of forming a component of an integrated multiregion model. An outline is given of the structure of interregional programming models in both economy-wide and individual sector applications. We then discussed:

(1) how inclusion of input-output relations and the results of comparative cost and industrial complex analyses as constraints in appropriately designed economy-wide programming models permitted

-levels of regional investment (capacity expansion) in each sector to be derived endogenously instead of being consolidated into an exogenously determined set of final demands, and

-levels of regional activity to be derived in a manner consistent with the policy goals (ambitions) of different sectors, regions and/or levels of government; and

(2) how the regional production/trade patterns for one sector (say energy or agriculture) could be projected via the use of an interregional programming model, for other sectors (say mining and mineral processing) by use of comparative cost and industrial complex analyses, and for the remaining sectors via the use of an interregional input-output model.

We concluded that in the Australian context a linear programming module would be most useful when applied to agricultural sectors in those regions where alternative land uses, crop rotations, etc. are possible.

Chapter 5 was concerned with the development of a demographic module for use in the Australian context. This module includes a set of econometric equations for determining regional out- and in- migration flows, and a doubly constrained gravity model for determining the pattern of interregional migration flows. A sequence of steps was presented for achieving a meaningful linkage with the input-output, comparative cost-industrial complex and linear programming modules developed earlier. It was argued that the operation of such an integrated multiregion model would permit superior economic and demographic projections to be made.

Chapter 6 dealt with the development of a multiregional econometric module capable of implementation in the Australian context. For each region this module includes equations for determining household income by type, government expenditures by type by level, private investment expenditure by sector, and employment demand, labour supply and wage rates by demographic class. We then discussed how the projections derived from operating such an econometric module could be used to revise the final demand vector employed when operating an interregional input-output module. This then allowed the development of a sequence of steps which may be used to link multiregional input-output, comparative cost-industrial complex, linear programming, demographic and econometric modules -- and thereby to operate an integrated model containing these modules as its components.

In Chapter 7, we described a factor demand (substitution) module which may be added to the integrated model discussed earlier. In particular, two alternative functional forms -- the translog and Diewert (generalized Leontief) -- were considered for use in projecting substitution patterns between factor inputs (capital, labour, energy and materials) given the corresponding factor price changes. We then developed a number of procedures which could be used to revise the technical-trade coefficients comprising the interregional input-output module. In the Australian context such a module can only be made operational for manufacturing sectors, however its inclusion for even this limited set of sectors is felt to enhance the quality of projections which can be made using the integrated model.

Chapter 8 considers the potential extension of the model to include a national econometric module and a transportation module. Inclusion of such modules brings into sharper focus the problem of achieving top down-bottom up consistency in output, employment, population, etc. projections. Some preliminary remarks are made concerning this problem, as well as those concerned with the development of a solution algorithm for the overall model, however additional research is required here.

Finally, in Chapter 9 we point up several key policy areas for Australia and then identify several approaches which could be adopted to analyse policy initiatives in these areas using the integrated model. It is argued that this explicit consideration of the policy variable is critically important since not only does it enable us to make more accurate economic-demographic projections for each region comprising the system, but it also could enable us to feed the outputs derived from running our integrated model back into the policy formation process.

The objective of this book has been to develop an operational model that is much more comprehensive than those currently in use, in the sense that many diverse modules are effectively linked together. However, it should be recognized that for some questions a less comprehensive model may be more desirable -- especially when resources are scarce. At the same time it is clear that for other purposes additional interconnections and linkages should be effected, especially as regards:

(1) connecting the multiregion system of Australia with the world economy; and

(2) developing a module capable of exploring income distribution effects such as would be derived from a microsimulation approach.

Although beyond the scope of this book, research is clearly needed in these areas as well as those mentioned in Chapter 8.

REFERENCES

Adams, F.G. and N.J. Glickman (1980) "Perspectives on Multiregional Modelling" in Adams, F.G. and N.J. Glickman (eds) Modelling the Multiregion Economic System (Lexington Books, Lexington, MA.).

Ahlburg, D.A. (1982) "Forecasting Regional Births from National Birth Forecasts" Paper presented at International Conference on Forecasting Regional Population Change and Its Economic Determinants and Consequences, Airlie, Virginia, May 26-29.

Airov, J. (1956) "Location Factors in Synthetic Fiber Production" Papers, Regional Science Association, Vol. 2, pp. 291-303.

Airov, J. (1959) The Location of the Synthetic Fiber Industry: A Study in Regional Analysis (Wiley, New York).

Albegov, M. and A. Umnov (1981) "Linkage of Regional Models" Working Paper,WP-81-85, International Institute for Applied Systems Analysis, Laxenburg, Austria, June.

Albegov, M., A.E. Andersson and F. Snickars (eds.) (1982) Regional Development Modelling: Theory and Practice (North Holland, Amsterdam).

Allen, R.R., L.R. Babcock, and N.L. Najda (1975) "Air Pollution Dispersion Modelling: Application and Uncertainty" Regional Science Perspectives, Vol. 5, pp. 1-26.

Almon, C., D. Belzer and P. Taylor (1979) "Prices in Input-Output: The INFORUM Experience in Modelling the U.S. Deregulation of Domestic Oil" Journal of Policy Modelling, Vol. 1, No. 3, pp. 399-412.

Almon, C., M.D. Buckler, L.R. Horwitz and T.C. Reimbold (1974) 1985: Interindustry Forecasts of the American Economy (Lexington Books, Lexington, MA.).

Alonso, W. (1977) "Policy-Oriented Interregional Demographic Accounting and a Generalization of Population Flow Models" in Brown, A.A. and E. Neuberger (eds) International Migration: A Comparative Perspective (Academic Press, New York).

Alperovich, G., J. Bergsman, and C. Ehemann (1977) "An Econometric Model of Migration between U.S. Metropolitan Areas" Urban Studies, Vol. 14, pp. 135-145.

Andersson, A.E. (1975) "A Closed Nonlinear Growth Model for International and Interregional Trade and Location" Regional Science and Urban Economics, Vol. 5, No. 4, pp. 427-444.

REFERENCES

Andersson, A.E. and D.F. Batten (1979) "An Interdependent Framework for Integrated Sectoral and Regional Development" *Working Paper, WP-79-97,* International Institute for Applied Systems Analysis, Laxenburg, Austria, October.

Andersson, A.E. and D. Philipov (1982) "Economic Models of Migration" in M. Albegov, et al. (eds) *Regional Development Modelling: Theory and Practice* (North-Holland, Amsterdam), pp. 105-25.

Anselin, L. and W. Isard (1979) "On Alonso's General Theory of Movement" *Man, Environment, Space and Time,* Vol. 1, No. 1, pp. 52-63.

Arntzen, J.W., L.C. Braat, and L. Hordijk (1980) "An Integrated Environmental Model for Regional Policy Analysis" Paper presented at First World Regional Science Congress, Cambridge, Mass., June.

Ballard, K.P. and N.J. Glickman (1977) "A Multiregional Econometric Forecasting System: A Model for the Delaware Valley" *Journal of Regional Science*, Vol. 17, No. 2, pp. 161-177.

Ballard, K.P., N.J. Glickman, and R.D. Gustely (1980) "A Bottom-up Approach to Multiregional Modelling: NRIES" in Adams, F.G. and N.J. Glickman (eds) *Modelling the Multiregion Economic System* (Lexington Books, Lexington, MA.).

Ballard, K.P., N.J. Glickman and R.M. Wendling (1980) "Using a Multiregional Econometric Model to Measure the Spatial Impact of Federal Policies" in Glickman, N.J. (ed.) *The Urban Impacts of Federal Policies* (John Hopkins Univ. Press, Baltimore, MD), pp. 192-216.

Ballard, K.P., R.D. Gustely, and R.M. Wendling (1980) *National-Regional Impact Evaluation System: Structure Performance and Applications of a Bottom-up Interregional Econometric Model* (United States Department of Commerce, Washington, D.C.).

Ballard, J. and R. Wendling (1980) "The National-Regional Impact Evaluation System: A Spatial Model of U.S. Economic and Demographic Activity" *Journal of Regional Science,* Vol. 20, No. 2, pp. 143-158.

Baranov, E.F. and I.S. Matlin, 1982, "A System of Models for Co-ordinating Sectoral and Regional Development Plans" in Issaev, B., et al. (ed) *Multiregional Economic Modelling: Practice and Prospect* (North Holland, Amsterdam), pp. 143-155.

Baranov, E.F., I.S. Matlin and A.V. Koltsov (1980) "Multiregional and Regional Modelling in the USSR" in F.G. Adams and N.J. Glickman (eds) *Modelling the Multiregional Economic System* (Lexington Books, Lexington, MA), pp. 215-220.

REFERENCES

Bargur, J. (1969) "A Dynamic Interregional Input-Output Programming Model of the California and Western States Economy" Contribution No. 128 (Water Resources Centre, California, June).

Barnard, J.R., J.A. MacMillan and W.R. Maki (1969) "Evaluation Models for Regional Development Planning" Papers, Regional Science Association, Vol. 23, pp. 117-138.

Batten, D.F. (1981) "Entropy, Information Theory, and Spatial Input-Output Analysis" Umea Economic Studies Series, No. 92, University of Umea, Sweden, June.

Batten, D.F. (1981a) "Toward a System of Models for Integrated National and Regional Development" Umea Economic Studies Series, No. 93, University of Umea, Sweden, May.

Batten, D.F. (1982) "The Interregional Linkages between National and Regional Input-Output Models" International Regional Science Review, Vol. 7, No. 1, pp. 53-67.

Batten, D.F. and A.E. Andersson (1981) "Towards a Hierarchial System of Models to Co-ordinate National and Regional Economic Developments" Paper presented at the Seventh Pacific Regional Science Conference, Surfers' Paradise, Australia, August.

Batty, M. (1976) Urban Modelling (Cambridge University Press, New York).

Batty, M. and S. Mackie (1972) "The Calibration of Gravity, Entropy and Related Models of Spatial Interaction" Environment and Planning, Vol. 4, pp. 131-150.

Beaumont, P., A. Isserman, D. McMillen, D. Plane and P. Rogerson (1982) "The ECESIS Economic-Demographic Interregional Model of the United States" Paper presented at International Conference on Forecasting Regional Population Change and Its Economic Determinants and Consequences, Airlie, Virginia, May 26-29.

Bergman, L. and A. Por (1980) "A Quantitative General Equilibrium Model of the Swedish Economy" Working Paper, WP-80-4, International Institute of Applied Systems Analysis, Laxenburg, Austria, October.

Berndt, E.R. and L.R. Christensen (1973) "The Internal Structure of Functional Relationships: Separability, Substitution and Aggregation" Review of Economic Studies, Vol. 40, pp. 402-410.

Berndt, E.R. and D.O. Wood (1975) "Technology, Prices and the Derived Demand for Energy" Review of Economics and Statistics, Vol. 57, No. 3, pp. 259-268.

Beston, D. (1981) "Complementary Strategies for Policy Analysis: Combining Microeconomic and Regional Simulation Models" Paper presented at International Conference on Structural Economic Analysis and Planning in Time and Space, Umea, Sweden, June.

Beyers, W.B. (1978) "On the Structure and Development of Multiregional Economic Systems" Papers, Regional Science Association, Vol. 40, pp. 109-133.

Beyers, W.B. (1980) "Migration and the Development of the Multiregional Economic System" Economic Geography, Vol. 56, No. 4, pp. 320-334.

Billings, R. (1969) "The Mathematical Identity of the Multipliers Derived from the Economic Base Model and the Input-Output Model" Journal of Regional Science, Vol. 9, pp. 471-473.

Birg, H. (1981) "An Interregional Population-Employment Model for the Federal Republic of Germany: Methodology and Forecasting Results for the Year 2000" Papers, Regional Science Association, Vol. 47, pp. 97-117.

Black, W. (1972) "Interregional Commodity Flows: Some Experiments with the Gravity Model" Journal of Regional Science, Vol. 12, No. 1, pp. 107-118.

Bolton, R. (1980a) "Multiregional Models: An Introduction to a Symposium" Journal of Regional Science, Vol. 20, No. 2, pp. 131-142.

Bolton, R. (1980b) "Multiregional Models for Policy Analysis" in Adams, F.G. and N.J. Glickman (eds) Modelling the Multiregional Economic System (Lexington Books, Lexington, MA.).

Bolton, R. (1982a) "Industrial and Regional Policy in Multiregion Modelling" in Bell, M.E. and P.S. Lande (eds) Regional Dimensions of Industrial Policy (Lexington Books, Lexington, MA).

Bolton, R. (1982b) "The Development of Multiregional Economic Modelling in North America: Multiregional Models in Transition for Economies in Transition" in Issaev, B., et al (eds) Multiregional Economic Modelling: Practice and Prospect (North Holland, Amsterdam), pp. 157-170.

Bourque, P., et al. (1967) The Washington Economy: An Input-Output Study (Graduate School of Business Administration, University of Washington, Seattle, WA.).

Boyce, D. F. and G.D. Hewings (1980) "Interregional Commodity Flow, Input-Output and Transportation Modelling: An Entropy Formulation" Paper presented at First World Regional Science Congress, Cambridge, Mass., June.

Bronzini, M.S., J.H. Herendeen, J.H. Miller and N.K. Womer (1974) "A Transportation-Sensitive Model of a Regional Economy" Transportation Research, Vol. 8, pp. 45-62.

Brown, M., M. di Palma, and B. Ferrara (1972) "A Regional-National Econometric Model of Italy" Papers, Regional Science Association, Vol. 19, pp. 25-44.

Brown, M., M. di Palma and B. Ferrara (1972) Regional-National Econometric Modelling -- with an Application to be Italian Economy (Pion, London).

Brownrigg, M. and M. Greig (1975) "Differential Multipliers for Tourism" Scottish Journal of Political Economy, Vol. 22, pp. 261-275.

Bureau of Mineral Resources (1979) The Australian Mineral Industry Annual Review (Australian Government Publishing Service, Canberra, A.C.T.).

Bureau of Mineral Resources (1981) Australian Mineral Industry Quarterly Review (Australian Government Publishing Service, Canberra, A.C.T.).

Butz, W. and M. Ward (1979) "Will U.S. Fertility Remain Low?: A New Economic Interpretation" Population and Development Review, Vol. 5, No. 4, pp. 663-688.

Caldwell, S. (1980) "Individual and Household Behavior: An Interregional Microanalytic Model" Unpublished Manuscript, Department of Sociology, Cornell University.

Caldwell, S. (1982) "Modelling Demographic-Economic Interactions: Micro, Macro and Micro/Macro Strategies" Paper presented at International Conference on Forecasting Population Change and Its Economic Determinants and Consequences, Airlie, Virginia, May 26-29.

Caldwell, S., W. Greene, T. Mount, S. Saltzman and R. Broyd (1979) "Forecasting Regional Energy Demand with Linked Micro/Macro Models" Papers, Regional Science Association, Vol. 43, pp. 99-113.

Caldwell, S. and S. Saltzman (1981) "Microsimulation and Regional Science" Paper presented at North American Meetings of Regional Science Association, Montreal, Canada, November.

Carrothers, G.A.P. (1956) "A Historical Review of the Gravity and Potential Concepts of Human Interaction" Journal of American Institute of Planners, Vol. 22, pp. 94-102.

Carrothers, G.A.P. (1959) "Forecasting the Population of Open Areas," Unpublished Ph.D. dissertation, Massachusetts Institute of Technology, Cambridge, Mass.

Cebula, R.J. (1979) <u>The Determinants of Human Migration</u> (Lexington Books, Lexington, MA.).

Chenery, H. (19653) "Regional Analysis" in Chenery, H., P. Clark, and C. Pinna (eds) <u>The Structure and Growth of the Italian Economy</u> (US Mutual Security Agency, Rome), pp. 96-115.

Chenery, H. and P. Clark (1959) <u>Interindustry Economics</u> (Wiley, New York).

Cherniavsky, E. (1974) <u>Brookhaven Energy System Optimisation Model (BESOM)</u> (Brookhaven National Laboratory Research Report, BNL 19569, Upton, NY).

Christensen, L.R. and W.H. Greene (1976) "Economies of Scale in U.S. Electric Power Generation" <u>Journal of Political Economy,</u> Vol 84, pp. 655-676.

Christensen, L.R. and L.J. Lau (1971) "Conjugate Duality and the Transcendental Logarithmic Utility Function" <u>Econometrica,</u> Vol. 39, No. 4, pp. 255-256.

Christensen, L.R., D.W. Jorgenson and L.J. Lau (1973) "Transcendental Logarithmic Production Frontiers", <u>Review of Economics and Statistics,</u> Vol. 55, pp. 28-46.

Clapp, J. (1977) "The Relationships between Regional Input-Output, Intersectoral Flows and Rows - Only Analysis" <u>International Regional Science Review</u>, Vol. 2, No. 1, pp. 29-89.

Codsi, G. and K.R. Pearson (1988), "Developing and Implementing Large Economic Models Using GEMPACK, A General Purpose Software Suite" Preliminary Working Paper No IP-39, Impact Research Centre, University of Melbourne.

Coupe, B. (1977) <u>Regional Economic Structure and Environmental Pollution</u> (Martinus Nijhoff, Boston, MA).

Courbis, R. (1972) "The REGINA Model, a Regional-National Model of the French Economy" <u>Economics and Planning,</u> Vol. 12, No. 3, pp. 133-152.

Courbis, R. (1975) "Urban Analysis in the Regional-National Model REGINA of the French Economy" <u>Environment and Planning A</u>, Vol. 7, no. 7, pp. 873-878.

Courbis, R. (1979) "The REGINA Model, A Regional-National Model for French Planning" <u>Regional Science and Urban Economics,</u> Vol. 9, No. 2-3, pp. 117-139.

Courbis, R. (1980) "Multiregional Modelling and the Interaction between Regional and National Development: A General Theoretical Framework" in Adams, F.G. and N.J. Glickman (eds) Modelling the Multiregional Economic System (Lexington Books, Lexington, MA), pp. 107-130.

Courbis, R. (1982a) "Measuring Effects of French Regional Policy by Means of a Regional-National Model" Regional Science and Urban Economics, Vol. 21, No. 1, pp. 1-21.

Courbis, R. (1982b) "Multiregional Modelling: A General Appraisal" in Albegov, M., et al. (eds) Regional Development Modelling: Theory and Practice (North-Holland, Amsterdam), pp. 65-84.

Courbis, R. and D. Vallet (1976) "An Interindustry Interregional Table for the French Economy" in Polenske, K.R. and J.V. Skolka (eds) Advances in Input-Output Analysis (Ballinger, Cambridge, Ma), pp. 231-249.

Crow, R.T. (1973) "A Nationally-Linked Regional Econometric Model" Journal of Regional Science, Vol. 13, No. 2, pp. 187-204.

Crow, R.T. (1979) "Output Determination and Investment Specification in Macroeconomic Models of Open Regions" Regional Science and Urban Economics, Vol. 9, Nos. 2-3, pp. 141-158.

Cumberland, J. (1966) "A Regional Interindustry Model for Analysis of Development Objectives" Papers, Regional Science Association, Vol. 17, pp. 65-94.

Cumberland, J. and R. Korbach (1973) "A Regional Inter-industry Environmental Model" Papers, Regional Science Association, Vol. 30, pp. 61-75.

Cumberland, J. and B. Stram (1976) "Empirical Applications of Input-Output Models to Environmental Problems" in Polenske, K. and K. Skolka (eds) Advances in Input-Output Analysis (Ballinger, Cambridge, MA).

Czamanski, S. and L. Ablas (1979) "Identification of Industrial Clusters and Complexes: A Comparison of Methods and Findings" Urban Studies, Vol. 16, pp. 61-80.

Czamanski, S. and D. Czamanski (1976) Study of Formation of Spatial Complexes (Dalhousie University, Halifax, Canada).

Daly, H. (1968) "On Economics as a Life Science" Journal of Political Economy, Vol. 76, pp. 392-406.

Dantzig, G. (1976) "On the Reduction of an Integrated Energy and Interindustry Model to a Smaller Linear Program" Review of Economic Studies, Vol. 58, No. 2, pp. 248-250.

David, P.A. (1974) "Fortune, Risk and the Microeconomics of Migration" in David, P.A. and M.W. Reder (eds) Nations and Households in Economic Growth (Academic Press, New York).

Denny, M. and M. Fuss (1977) "The Use of Approximation Analysis to Test for Separability and the Existence of Consistent Aggregates" American Economic Review, Vol. 67, pp. 404-418.

Diewert, W. (1971) "An Application of the Shephard Duality Theorem: Generalized Leontief Production Function" Journal of Political Economy, Vol. 79, pp. 481-507.

Diewert, W. (1974) "Applications of Duality Theory" in Intriligator, M. and D.A. Kendrick (eds) Frontiers in Quantitative Economics, Vol. II (North Holland, Amsterdam), pp. 106-170.

Diewert, W. (1976) "Exact and Superlative Index Numbers" Journal of Econometrics, Vol. 4, pp. 115-146.

Dixon, P. B. (1979) "A Skeletal Version of ORANI '78: Theory, Data and Application" Preliminary Working Paper, No. OP-24, IMPACT Project, Melbourne, June.

Dolenc, M. (1968) "The Bucks County Interregional Input-Output Study" Papers, Regional Science Association, Vol. 20, pp. 43-54.

Domenich, T. and D. McFadden (1975) Urban Travel Demand: A Behavioural Analysis (North Holland, Amsterdam).

Don, F.J. and G. M. Gallo (1987) "Solving Large Sparse Systems of Demand Equations in Econometric Models" Journal of Forecasting, Vol. 6, pp. 167-180.

Dorfman, R., P.A. Samuelson and R. Solow (1958) Linear Programming and Economic Analysis (McGraw Hill, New York).

Dresch, S. (1980) "IDIOM: A Sectoral Model of the National and Regional Economies" in Adams, F.G. and N.J. Glickman (eds) Modelling the Multiregional Economic System (Lexington Books, Lexington, MA), pp. 161-165.

Dresch, S. and D. Updegrove (1980) "IDIOM: A Disaggregated Policy-Impact Model of the US Economy" in R.H. Haveman and K. Hollenbeck (eds) Microeconomic Simulation Models for Public Policy Analysis, Vol. 2 (Academic Press, New York), pp. 213-249.

Drud, A. (1983) "A Survey of Model Representations and Simulation Algorithms in Some Existing Modelling Systems" <u>Journal of Economic Dynamics and Control</u>, Vol. 5, pp. 5-35.

Drud, A. (1983a) "Interfacing Modelling Systems and Solution Algorithms" <u>Journal of Economic Dynamics and Control</u>, Vol. 5, pp. 131-149.

Dulgeroff, J.E. (1980) "An Interregional Demoeconomic Simulation Model of Population Distribution for Arizona and California" Paper presented at North American Meetings of the Regional Science Association, Milwaukee, November.

Edwards, G. (1977) "An Input-Output Model of the Tasmanian Economy" Paper presented at the Second Regional Science Conference (Australian and New Zealand Section), University of New South Wales, Sydney, December.

Emerson, M.J. and T.L. Hunt (1980) "Regional Differences in Post-Embargo Capital-Labour-Energy Substitution in U.S. Manufacturing" Paper presented at First World Regional Science Congress, Cambridge, Mass., June.

Erlander, S. (1977) "Accessibility, Entropy and the Distribution and Assignment of Traffic" <u>Transportation Research</u>, Vol. 11, pp. 149-153.

Eskelinen, H. (1981) "Core and Periphery in a Three Region Input-Output Framework" Paper presented at the European Meeting of the Regional Science Association, Barcelona, Spain, August.

Evans, A.W. (1971) "The Calibration of Trip Distribution Models with Exponential or Similar Cost Functions" <u>Transportation Research,</u> Vol. 5, pp. 15-38.

Evans, M. and J. Baxter (1980) "Regionalising National Projections with a Multiregional Input-Output Model Linked to a Demographic Model" <u>Annals of Regional Science</u>, Vol. 14, pp. 52-71.

Falleur, R., De, H. Bogaert, T. de Biolley, and P.Huge (1975) "Use of the RENA Model for Forecasting the Main Lines of Medium-Term Economic Policy" in Economic Commission for Europe (ed) <u>Use of Systems of Models in Planning</u> (United Nations, New York).

Feder, G. (1979) "Alternative Opportunities and Migration: An Exposition" <u>Annals of Regional Science</u>, Vol. 13, No. 1, pp. 57-67.

Feder, G. (1980) "Alternative Opportunities and Migration: Evidence from Korea" <u>Annals of Regional Science</u>, Vol. 14, No. 1, pp. 1-11.

Feeny, G. (1973) "Two Models of Multiregional Population Dynamics" <u>Environment and Planning A,</u> Vol. 5, No. 1, pp. 31-43.

Field, B.C. and C. Grebenstein (1980) "Capital-Energy Substitution in U.S. Manufacturing" Review of Economics and Statistics, Vol. 62, pp. 207-212.

Fields, G. (1976) "Labour Force Migration, Unemployment and Job Turnover" Review of Economic Statistics, Vol. 58, pp. 407-415.

Fjeldsted, B.L. and J.B. South (1979) "A Note on the Multiregion Multiindustry Forecasting Model" Journal of Regional Science, Vol. 19, No. 4, pp.483-491.

Folmer, H. (1980) "Measurement of the Effects of Regional Policy Instruments", Environment and Planning A, Vol. 12, pp. 1191-1202.

Fromm, D., J. Hill, C. Loxely and M. McCarthy (1979) "The Wharton-EFA Multiregional Econometric Model: A Bottom-Up Approach" Paper presented at the Conference on Multiregional Models, First World Regional Science Congress, Cambridge, Mass., June.

Fromm, D., C. Loxely and M. McCarthy (1980) "The Wharton EFA Multiregional Econometric Model: A Bottom-Up/Top-Down Approach to Constructing a Regionalised Model of a National Economy" in Adams, F.G. and Glickman N.J. (eds) Modelling the Multiregional Economic System (Lexington Books, Lexington, MA), pp. 89-106.

Fuhrman, W. (1980) "The International Dependencies of Small Open Economies" Paper presented at International Conference on Structural Economic Analysis and Planning in Time and Space, Umea, Sweden, June.

Funck, R. and G. Rembold (1975) "A Multiregional , Multisectoral Forcasting Model for the Federal Republic of Germany" Papers, Regional Science Assocation, Vol. 34, pp. 69-82.

Fuss, M. (1977) "The Demand for Energy in Canadian Manufacturing" Journal of Econometrics, Vol. 5, pp. 89-116.

Fuss, M. and D. McFadden (eds) (1978) Production Economics: a Dual Approach to Theory and Applications, Vol. 1 and 2, (North Holland, Amsterdam).

Fuss, M., D. McFadden and Y. Mundlak (1978) "A Survey of Functional Forms in the Economic Analysis of Production" in Fuss, M., and D. McFadden (eds) Production Economics: A Dual Approach to Theory and Applications (North Holland, Amsterdam).

Garnick, D. (1969) "Disaggregated Basic-Service Models and Regional Input-Output Models in Multiregional Projections" Journal of Regional Science, Vol. 9, pp. 87-99.

Garnick, D. (1970) "Differential Regional Multiplier Models" Journal of Regional Science, Vol.10, pp. 35-47.

Gerking, S. (1977) "Reconciling 'Rows Only' and 'Columns Only' Coefficients in an Input-Output Model" International Regional Science Review, Vol. 1, No. 2, pp. 30-46.

Giarratani, F. (1974) "Air Pollution Abatement: Output and Relative Price Effects, a Regional Input-Output Simulation" Environment and Planning A, Vol. 6, pp. 307-312.

Giarratani, F. (1980) "A Note on a Neglected Aspect of Intersectoral Flows Analysis" Journal of Regional Science, Vol. 20, No. 4, pp. 513-515.

Gillen, W.J. and Guccione A. (1980) "International Feedbacks in Input-Output Models: Some Formal Results" Journal of Regional Science, Vol. 20, No. 4, pp. 477-482.

Glejser, H., G. van Daele and M. Lambrecht (1973) "First Experiments with an Econometric Model of the Belgian Economy" Regional Science and Urban Economics, Vol. 3, No. 3, pp. 301-314.

Glickman, N.J. (1974) "Son of the Specification of Regional Econometric Models" Papers, Regional Science Association, Vol. 32, pp. 155-174.

Glickman, N.J. (1982) "Using Empirical Models for Regional Policy Analysis" in Albegov, M., et al. (eds) Regional Development Modelling: Theory and Practice (North-Holland, Amsterdam), pp. 85-104.

Gober-Meyers, P. (1978) "Interstate Migration and Economic Growth: A Simultaneous Equations Approach" Environment and Planning A, Vol. 10, No. 11, pp. 1241-1252.

Goettle, R. (1977) "Regional Energy System Implications of Alternative Coal Conversion/Utilisation Policies: Methodology" Paper presented at North American Meetings of the Regional Science Association, Philadelphia, November.

Goettle, R., E. Cherniavsky, and R. Tessmer (1977) An Integrated Multiregional Energy and Interindustry Model of the United States (Brookhaven National Laboratory Research Report, BNL 22728, Upton, NY).

Golladay, F. and R. Haveman (1977) The Economic Impacts of a Tax Transfer Policy (Academic Press, New York).

Gordon, I.R. (1974) "A Gravity Flows Approach to an Interregional Input-Output Model of the UK" in Cripps, E.L. (ed) Space-Time Concepts in Urban and Regional Models (Pion Press, London), pp. 56-73.

Gordon, I.R. (1977) "Regional Interdependence in the United Kingdom Economy" in Leontief, W.W. (ed) Structure, System and Economic Policy, (Cambridge University Press, England), pp. 111-122.

Gordon, I.R. (1978) "Distance Deterrence and Commodity Values" Environment and Planning A, Vol. 10, No. 7, pp. 889-900.

Gordon, I.R. (1979) "Freight Distribution Models Compared: A Comment" Environment and Planning A, Vol. 11, No. 2, pp. 219-221.

Gordon, I.R. and D. Pitfield (1981) "A Multistream Approach to the Analysis of Hierarchically Differentiated Spatial Interactions" Paper presented at European Meetings of the Regional Science Assocaition, Barcelona, Spain, August.

Gordon, P. and J. Ledent (1981) "Towards an Interregional Demoeconomic Model" Journal of Regional Science, Vol. 21, No. 1, pp. 79-87.

Gordon, P. and J. Ledent (1980) "Modelling the Dynamics of a System of Metropolitan Areas: A Demoeconomic Approach" Environment and Planning A, Vol. 12, No. 2, pp. 125-134.

Granberg, A. (1982) "Experience in the Use of Multiregional Economic Models in the Soviet Union" in Issaev, B., et al. (eds) Multiregional Economic Modelling: Practice and Prospect (North-Holland, Amsterdam) pp. 135-142.

Granholm, A. (1981) Interregional Planning Models for the Allocation of Private and Public Investments (University of Gothenburg, Department of Economics, Gothenburg).

Granholm, A. and F. Snickars (1979) "An Interregional Planning Model of Private and Public Investment Allocation" Paper presented at European Meeting of the Regional Science Association, London, August.

Greenwood, M.J. (1975) "Research on Internal Migration in the United States: A Survey" Journal of Economic Literature, Vol. 13, No. 2, pp. 397-433.

Greenwood, M.J. (1981) Migration and Economic Growth in the United States (Academic Press, New York).

Gregory, P., J. Campbell and B. Cheng (1972) "A Simultaneous Equation Model of Birth Rates in the United States" Review of Economics and Statistics, Vol. 49, No. 4, pp. 374-380.

Greytak, D. (1969) "A Statistical Analysis for Regional Export Estimating Techniques" Journal of Regional Science, Vol. 9, pp. 387-395.

Greytak, D. (1970) "Regional Impact of Interregional Trade in Input-Output Analysis" Papers, Regional Science Association, Vol. 25, pp. 203-217.

Griffin, J.M. and P.R. Gregory (1976) "An Intercountry Translog Model of Energy Substitution Responses" American Economic Review, Vol. 66, pp. 845-857.

Halfkamp, W. and P. Nijkamp (1980) "An Integrated Interregional Model for Pollution Control" in Lakshmanan, T.R. and P. Nijkamp (eds) Economic-Environmental-Energy Interaction Modelling and Policy Analysis (Martinus Nijhoff, Boston, MA).

Halvorson, R. (1977) "Energy Substitution in U.S. Manufacturing" Review of Economics and Statistics, Vol. 59, pp. 381-388.

Halvorson, R. and J. Ford (1978) "Substitution Among Energy, Capital and Labour Inputs in U.S. Manufacturing" in Pindyck, R.S. (ed), Advances in Economics of Energy and Resources, Vol. 1 (J.A.I. Press, Greenwich, CN).

Hansen, W. and C. Tiebout (1963) "An Intersectoral Flows Analysis of the California Economy" Review of Economics and Statistics, Vol. 45, pp. 409-418.

Harris, C.C. (1970) "A Multiregional Multiindustry Forecasting Model" Papers, Regional Science Association, Vol. 25, pp. 169-180.

Harris, C.C. (1973) The Urban Economies, 1985: A Multiregional Multiindustry Forecasting Model (Lexington Books, Lexington, MA).

Harris, C.C. (1974) Regional Economic Effects of Alternative Highway Systems (Ballinger, Cambrdige, MA).

Harris, C.C. (1980) "The Multiregional Multiindustry Forecasting Model" in Pleeter, S. (ed) Economic Impact Analysis: Methodology and Applications (Martinus Nijhoff, Boston, MA.).

Harris, C.C. (1980a) "New Developments and Extensions of the Multiregional, Multiindustry Forecasting Model" Journal of Regional Science, Vol. 20, No. 2, pp. 159-171.

Harris, C.C. and M. Nadji (1980) "The Framework of the Multiregional Multiindustry Forecasting Model" in Adams, F.G. and N.J. Glickman (eds) Modelling the Multiregion Economic System (Lexington Books, Lexington, MA.).

Hart, R.A. (1975) "Interregional Economic Migration: Some Theoretical Considerations, Part I" Journal of Regional Science, Vol. 15, No. 2, pp. 127-138.

Hart, R.A. (1975a) "Interregional Economic Migration: Some Theoretical Considerations, Part 2" *Journal of Regional Science*, Vol. 15, No. 3, pp. 189-305.

Hartwick, H. (1971) "Notes on the Isard and Chenery-Moses Interregional Input-Output Models" *Journal of Regional Science*, Vol. 11, No. 1, pp. 73-86.

Hashim, S.R. and P.N. Mathur (1975) "Interregional Programming Models for Economic Development" in Cripps, E.L. (ed) *Regional Science-New Concepts and Old Problems* (Pion, Ltd., London).

Haveman, R. and K. Hollenbeck (1980) *Microeconomic Simulation Models for the Analysis of Public Policy*, Volumes I and II, (Academic Press, New York).

Heady, E.O. (1964) "Discussion of Some Particular Interregional Programming Models" *Papers, Regional Science Association*, Vol. 13, pp. 121-126.

Henderson, J. (1985) *The Efficiency of the Coal Industry: An Application of Linear Programming* (Harvard University Press, Cambridge, MA.).

Henderson, J.M. and R.E. Quandt (1980) *Microeconomic Theory: A Mathematical Approach* (3rd ed) (McGraw Hill, New York).

Higgs, P.J., B.R. Parmenter and R. Rimmer (1981) "Incorporating Regional Dimensions in Economy-Wide Models: A Preliminary Report on a Tasmanian Version of ORANI" Paper presented to the Seventh Pacific Regional Science Conference, Surfers Paradise, Australia, August.

Hite, J. and E. Laurent (1972) *Environmental Planning: An Economic Analysis* (Praeger, New York).

Hodge, G. and C.C. Wong (1972) "Adapting Industrial Complex Analysis to the Realities of Regional Data" *Papers, Regional Science Association*, Vol. 28, pp. 145-166.

Hollenbeck, K. (1980) "Linking Microeconomic Simulation Estimates of Household Final Demand to a Multiregional Model of the Economy" in Adams, F.G. and N.J. Glickman (eds) *Modelling the Multiregional Economic System* (Lexington Books, Lexington, MA).

Hoyt, H. (1949) *The Economic Base of the Brockton, Massachusetts Area* (Brockton, MA.).

Hua, C.I. and F. Porcell (1979) "A Critical Review of the Development of the Gravity Model" *International Regional Science Review*, Vol. 4, No. 2, pp. 97-126.

REFERENCES

Hudson, E.A. and D.W. Jorgenson (1974) "U.S. Energy Policy and Economic Growth, 1975-2000" Bell Journal of Economics and Management Science, Vol. 5, pp. 461-514.

Hudson, E.A. and D.W. Jorgenson (1976) "Tax Policy and Energy Conservation" in Jorgenson, D.W. (ed) Econometric Studies of U.S. Energy Policy (North Holland, Amsterdam).

Hulten, C.R. (1973) "Divisa Index Numbers" Econometrica, Vol. 41, pp. 1017-1026.

Hurter, A. P. and L.N. Moses (1964) "Regional Investment and Interregional Programming" Papers, Regional Science Association, Vol. 14, pp. 105-120.

Hyman, G.M (1969) "The Calibration of Trip Distribution Models" Environment and Planning A, Vol. 1, pp. 105-112.

Isard, W. (1951) "Interregional and Regional Input-Output Analysis: A Model of a Space Economy" Review of Economics and Statistics, Vol. 33, pp. 318-328.

Isard, W. (1953) "Some Empirical Results and Problems of Regional Input-Output Analysis" in Leontief, W., et al Studies in the Structure of the American Economy (Oxford Univ. Press, New York).

Isard, W. (1956) Location and Space Economy (Wiley, New York).

Isard, W. (1958) "Introduction to Linear Programming" Journal of Regional Science, Vol. 1, No. , pp. 1-59.

Isard, W. (1960) "The Scope and Nature of Regional Science" Papers, Regional Science Association, Vol. 6, pp. 9-34.

Isard, W. (1969) "Some Notes on the Linkage of the Ecologic and Economic Systems" Papers, Regional Science Association, Vol. 22, pp. 85-96.

Isard, W. (1974) "Activity-Industrial Complex Analysis for Environmental Management" Papers, Regional Science Association, Vol. 33, pp. 128-140.

Isard, W. (1975) Introduction to Regional Science (Prentice Hall, Englewood Cliffs, NJ).

Isard, W. et al. (1960) Methods of Regional Analysis: An Introduction to Regional Science (MIT Press, Cambridge, MA).

Isard, W., et al. (1968) "On the Linkage of Socio-Economic and Ecologic Systems" Papers, Regional Science Association, Vol. 21, pp. 79-100.

Isard, W., et al. (1972) *Ecologic-Economic Analysis for Regional Development* (Free Press, New York).

Isard, W., et al. (1978) "Multiregional Input-Output: Materials for BNL Integrated Model" Unpublished Report, Field of Regional Science, Cornell University.

Isard, W. and L. Anselin (1978) "Crude Operational Multiregional Models for Identification of Integrated Interindustry, Employment and Environmental Affairs" Unpublished Report, Field of Regional Science, Cornell University.

Isard, W. and L. Anselin (1980) "Multiregional Comparative Cost-Industrial Complex, Input-Output, Programming Module" Paper presented at Conference on Multiregion Models, First World Regional Science Congress, Cambridge, Mass., June.

Isard, W. and L. Anselin (1982) "Integration of Multiregional Models for Policy Analysis" *Environment and Planning A*, Vol. 41, pp. 359-376.

Isard, W., D. Boyce, T.R. Lakshmanan, L.R. Klein, and S. B. Caldwell (1981) *Integration of Multiregional Models for Policy Analysis.* Research Proposal submitted to the US National Science Foundation, Vols, I, II and III, November.

Isard, W., F. Cesario and T. Reiner (1975) *Marginal Pollution Analysis for Long Range Forecasts.* Regional Science Dissertation and Monograph Series, No. 4 (P.U.R.S., Cornell University, Ithaca, NY).

Isard, W. and J. Cumberland (1950) "New England as a Possible Location for an Integrated Iron and Steel Works" *Economic Geography*, Vol. 26.

Isard, W. and C. Czamanski (1965) "Techniques for Estimating Local and Regional Multiplier Effects of Changes in the Level of Major Government Programs" *Papers, Peace Science Society International*, Vol. 3, pp. 19-46.

Isard, W. and T. Langford (1971) *Regional Input-Output Study: Recollections, Reflections and Diverse Notes on the Philadelphia Experience* (MIT Press, Cambridge, MA).

Isard, W. and L. Parcels (1977) "Regional Pattern of Iron and Steel Production 1985 and 2000" Unpublished Report, Field of Regional Science, Cornell Unversity.

Isard, W. and L. Parcels (1977a) "A Locational Analysis for the Petrochemical Industry" Unpublished Report, Field of Regional Science, Cornell University.

Isard, W. and L. Parcels (1977b) "A Locational Analysis for the Aluminium Industry" Unpublished Report, Field of Regional Science, Cornell University.

Isard, W. and E.W. Schooler (1955) <u>Location Factors in the Petrochemical Industry</u> (Office of Technical Services, U.S. Dept. of Commerce, Washington, D.C.).

Isard, W. and E.W. Schooler (1959) "Industrial Complex Analysis, Agglomeration Economies and Regional Development" <u>Journal of Regional Science,</u> Vol. 1, No. 2, pp. 19-33.

Isard, W., E.W. Schooler and T. Vietorisz (1959) <u>Industrial Complex Analysis and Regional Development</u> (Wiley, New York).

Isard, W. and C. Smith (1981) "Energy Policy Formulation with the Use of an Integrated Multiregion Model" Paper presented at Pacific Region Meetings of the Regional Science Association, Surfers Paradise, Australia, August.

Isard, W. and C. Smith (1982) <u>Conflict Analysis and Practical Conflict Management Procedures: An Introduction to Peace Science</u> (Ballinger Publishing Company, Cambridge, MA.).

Isard, W. and C. Smith (1983) "Linked Integrated Multiregion Models at the International Level", <u>Papers, Regional Science Association</u>, Vol. 51, pp. 3-20.

Isard, W. and C. Smith (1984a) "Incorporation of Conflict and Policy Analysis in an Integrated Multiregion Model' in A. Andersson, W. Isard and T. Puu (eds) <u>Regional and Industrial Development: Theories, Models and Empirical Evidence</u> (North-Holland, Amsterdam), pp. 109-31.

Isard, W. and C. Smith (1984b) "Policy Analysis Using an Integrated Multiregion Model" in M. Chatterji and P. Nijkamp (eds) <u>Spatial, Environment and Resource Policy in the Developing Countries</u> (Gower Publishing Company, Aldershot, UK), pp. 64-78.

Isard, W. and C. Smith (1984) "Fusion of the Abstract and Practical in Conflict Analysis and Management" in J.H.P. Paelinck and P.H. Vossen (eds) <u>Axiomatics and Pragmatics in Conflict Analysis</u> (Gower Publishing Company, Aldershot, UK).

Isard, W. and C. Smith (1986) "Estimating Demographic-Economic Linkages in an Integrated Multiregion Model" in A. Isserman (ed) <u>Population Change and the Economy: Social Science Theories and Models</u> (Kluwer-Nijhoff, Cambridge, MA.), pp. 159-75.

Isard, W., T. Smith, et al. (1969) <u>General Theory: Social, Political, Economic and Regional</u> (M.I.T. Press, Cambridge, MA.).

Isard, W. and T. Vietorisz (1955) "Industrial Complex Analysis and Regional Development, with Reference to Puerto Rico" *Papers, Regional Science Association,* Vol. 1, pp. U1-U17.

Isard, W. and V. Whitney (1952) *Atomic Power, an Economic and Social Analysis* (McGraw Hill, New York).

Isserman, A. M. (1977) "The Location Quotient Approach to Estimating Regional Economic Impacts" *Journal of American Institute of Planners,* Vol. 43, pp. 33-41.

Isserman, A.M. (1980) "ECESIS: An Economic-Demographic Forecasting Model of the States," Paper presented at American Statistical Association/U.S. Bureau of Census Evaluation Conference, Washington, November.

Isserman, A.M. (1982) "Multiregional Demoeconomic Modelling with Endogenously Determined Birth and Migration Rates: Theory and Prospects" Paper presented at International Conference on Forecasting Regional Population Change and Its Economic Determinants and Consequences, Airlie, Virginia, May 26-29.

Jensen, R.C. (1980) "The Concept of Accuracy in Regional Input-Output Models" *International Regional Science Review,* Vol. 5, pp. 139-154.

Jensen, R.C. and D. McGaurr (1976) "Reconciliation of Purchases and Sales Estimates in an Input-Output Table," *Urban Studies,* Vol.13, pp. 327-337.

Jensen, R.C. and G.R. West (1980) "The Effect of Relative Coefficient Size on Input-Output Multipliers" *Environment and Planning A,* Vol. 12, pp. 659-670.

Jensen, R.C., T.D. Mandeville and N.D. Karunaratne (1977) *Generation of Regional Input-Output Tables for Queensland* (Report to the Coordinator General's Dept. and the Dept. of Commercial and Industrial Development, Dept. of Economics, University of Queensland, St. Lucia).

Johnson, P.D., W.J. McKibben, and R.G. Trevor (1980) "Models and Multipliers" *Research Discussion Paper No. 8006,* Reserve Bank of Australia, Sydney, September.

Johnson, P.D., E.R. Moses and C.R. Wymer (1977) "The RBA 76 Model of the Australian Economy" *Conference on Applied Economic Research,* Reserve Bank of Australia, Sydney, December.

Johnston, J. (1972) *Econometrics* (McGraw Hill, New York).

Juster, F.T. and K.C. Land (eds) (1981) *Social Accounting Systems: Essays on the State of the Art* (Academic Press, New York).

Karasaka, G. (1969) "Manufacturing Linkages in the Philadelphia Economy: Some Evidence of External Agglomeration Forces" Geographical Analysis Vol. 1, pp. 354-369.

Karlqvist, A., R. Sharpe, D.F. Batten and J.F. Brotchie (1978) "A Regional Planning Model and Its Application to South Eastern Australia" Regional Science and Urban Economics, Vol. 8, pp. 57-86.

Kau, J.B. and Sirmans, C.F. (1979) "The Functional Form of the Gravity Model" International Regional Science Review, Vol. 4, No. 2, pp. 127-136.

Kau, J.B. and C.F. Sirmans (1979a) "A Recursive Model of Spatial Allocation of Migrants" Journal of Regional Science, Vol. 19, No. 1, pp. 47-56.

Klein, L.R. (1969) "The Specification of Regional Econometric Models" Papers, Regional Science Association, Vol. 23, pp. 105-115.

Klein, L.R., G. Adams, M. McCarthy, et al. (1980) "A Multiregional Econometric Module with a Bottom-Up Approach" Unpublished Manuscript, University of Pennsylvania.

Klein, L.R. and N.J. Glickman (1977) "Econometric Model-Building at the Regional Level" Regional Science and Urban Economics, Vol. 7, No. 1, pp. 3-23.

Kneese, A., R. Ayres, and R. D'Arge (1972) Economics and the Environment: A Materials Balance Approach (John Hopkins Press).

Lakshmanan, T.R. (1979) "A Multiregional Policy Model of the Economy, Environment and Energy" Working Paper, NSF -79-1, Boston University, Dept. of Geography.

Lakshmanan, T.R. (1980) "Multiregional Models of Factor Demand and Investment Supply" Working Paper, NSF-80-3, Boston University, Dept. of Geography.

Lakshmanan, T.R. (1982) "Integrated Multiregional Economic Modelling for the United States" in Issaev, B., et al. (eds) Multiregional Economic Modelling: Practice and Prospects (North Holland, Amsterdam), pp. 171-188.

Lakshmanan, T.R., W. Anderson and M. Jourabchi (1981) "Regional Dimensions of Factor and Fuel Substitution in U.S. Manufacturing" Regional Science and Urban Economics, Vol. 14, No. 3, pp. 381-398.

Lakshmanan, T.R., W. Anderson and M. Jourabchi and J. Chaubey (1980) "Factor Demand and Substitution in a Multiregional Framework" Paper presented at Conference on Multiregional Models, First World Regional Science Congress, Cambridge, Mass., June.

Lakshmanan, T.R. and S. Ratick (1980) "Integrated Models for Economic-Energy-Environmental Impact Analysis" in Lakshmanan, T.R. and P. Nijkamp (eds) <u>Economic-Environmental-Energy Interactions: Modelling and Policy Analysis</u> (Martinus Nijhoff, Boston, MA), pp. 7-39.

Ledent, J. (1978) "Regional Multiplier Analysis: A Demometric Analysis" <u>Environment and Planning A</u>, Vol. 10, pp. 537-560.

Ledent, J. (1982) "The Migration Component in Multiregional Economic-Demographic Forecasting Models: Using Alonso's Theory of Movement" Paper presented at International Conference on Forecasting Regional Population Change and Its Economic Determinants and Consequences, Airlie, Virginia, May 26-29.

Lee, D.B. (1973) "Requiem for Large-Scale Models," <u>Journal of the American Institute of Planners</u>, Vol. 39, pp. 163-178.

Lee, R. (1976) "Demographic Forecasting and the Easterlin Hypothesis" <u>Population and Development Review</u>, Vol. 2, pp. 459-468.

Lee, T.H., D.P. Lewis and J.R. Moore (1971) "Multiregion Intersectoral Flow Analysis" <u>Journal of Regional Science</u>, Vol. 11, No. 1, pp. 49-76.

Lee, T.H., J.R. Moore and D.P. Lewis (1973) <u>Regional and Interregional Intersectoral Flows Analysis</u> (University of Tennessee Press).

Leontief, W. (1953) "Interregional Theory" in Leontief, W.W., <u>et al. Studies in the Structure of the American Economy</u> (Oxford University Press, New York).

Leontief, W. (1953a) "Dynamic Analysis" in Leontief, W.W. <u>et al. Studies in the Structure of the American Economy</u> (Oxford University Press, New York), pp. 53-90.

Leontief, W. (1970) "Environmental Repurcussions and Economic Structure: An Input-Output Approach" <u>Review of Economics and Statistics</u>, Vol. 52, pp. 262-271.

Leontief, W. and D. Ford (1972) "Air Pollution and the Economic Structure: Empirical Results of Input-Output Computations" in Brody, A. and A. Carter (eds) <u>Input-Output Techniques</u> (North Holland, Amsterdam).

Leontief, W.W. and A. Strout (1963) "Multiregional Input-Output Analysis" in Barna, T. (ed) <u>Structural Interdependence and Economic Development</u> (Macmillian, New York), pp. 119-150.

Lesuis, P., F. Muller and P. Nijkamp (1980a) "Operational Methods for Stategic Environmental and Energy Policies" in Lakshmanan, T.R. and P. Nijkamp (eds) Economic-Environmental-Energy Interaction Modelling and Policy Analysis (Martinus Nijhoff, Boston, MA).

Lesuis, P., F. Muller and P. Nijkamp (1980b) "A Multiregional Conflict Model for Energy and Environmental Quality" Paper presented at European Meetings of the Regional Science Association, Munich, West Germany, August.

Lesuis, P., F. Muller and P. Nijkamp (1980c) "An Interregional Policy Model for Energy-Economic-Environmental Interactions" Regional Science and Urban Economics, Vol. 10, pp. 343-70.

Levy, M. and W. Wadycki (1974) "What is the Opportunity Cost of Moving?: Reconsideration of the Effects of Distance on Migration" Economic Development and Cultural Change, Vol. 22, pp. 198-214.

Lindsay, R. (1956) "Regional Advantage in Oil Refining" Papers, Regional Science Association, Vol. 2, pp. 304-317.

Liu, B.C. (1975) "Differential Net Migration Rates and the Quality of Life" Review of Economics and Statistics, Vol. 57, No. 3, pp. 329-337.

Lowry, I.S. (1966) Migration and Metropolitan Growth: Two Analytical Models (Chandler, San Francisco, CA).

Longmire, J.L., B.R. Brideoake, R.H. Blanks and N.H. Hall (1979) A Regional Programming Model of the Grazing Industry (Occasional Paper No. 48, Bureau of Agricultural Economics, Canberra).

Lundqvist, L. (1980) "A Dynamic Multiregional Input-Output Model for Analyzing Regional Development, Energy and Employment Use" Papers, Regional Science Association, Vol. 47, pp. 77-95.

Lundqvist, L. (1980a) "A Dynamic Multiregional Input-Output Model for Analysing Regional Development, Employment and Energy Use" Paper presented at European Meeting of Regional Science Association, Munich, West Germany, August.

Lundqvist, L. (1980b) "Energy Supply and Regional Development in Sweden: The Implications of Closing Down Nuclear Plants During the 80's" Paper presented at First World Regional Science Congress, Cambridge, Mass., June.

Lundqvist, L. (1981) "Multisectoral and Multiregional Analysis of a Small Open Economy - the Swedish Case" Paper presented at Conference on Structural Economic Analysis and Planning in Time and Space, Umea, Sweden, June.

Lundqvist, L. (1982) "Goals of Adaptivity and Robustness in Applied Regional and Urban Planning Models" in M. Albegov, et al (eds) <u>Regional Development Modelling: Theory and Practice</u> (North Holland, Amsterdam), pp. 185-203.

Maddala, G.S. (1977) <u>Econometrics</u> (McGraw Hill, New York).

Madden, M. and P. Batey (1980) "Achieving Consistency in Demographic-Economic Forecasting: Some Methodological Developments and Empirical Results" <u>Papers, Regional Science Association</u>, Vol. 44, pp. 91-106.

Madden, M. and P. Batey (1983) "Linked Population and Economic Models: Some Methodological Issues in Forecasting, Analysis and Policy Optimization" <u>Journal of Regional Science</u>, Vol. 23, No. 2, pp. 141-64.

Mandeville, T.D. (1976) <u>Linking APMAA to Representative Regional Input-Output Models</u> (APMAA Research Report #8, Department of Agricultural Economics and Business Management, University of New England).

Manne, A.S. (1985) "On the Formulation and Solution of Economic Equilibrium Models" <u>Mathematical Programming Study</u>, Vol. 23, pp. 1-22.

Mash, V.A. (1966) "The Multisectoral Interregional Long-Range Planning Problem" <u>Papers, Regional Science Association</u>, Vol. 18, pp. 87-90.

Mathur, P.N. (1972) "Multiregional Analysis in a Dynamic Input-Output Framework" in Carter, A.P. and A. Brody (eds) <u>Input-Output Techniques</u> (North Holland, Amsterdam).

McCarthy, M.D. (1972) "The Wharton Quarterly Econometric Forecasting Model, Mark III" <u>Studies in Quantitative Economics No. 6</u>, (Department of Economics, University of Pennsylvania, Philadelphia, PA).

McFadden, D. (1978) "Cost, Revenue and Profit Functions" in Fuss, M. and D. McFadden (eds) <u>Production Economics: A Dual Approach to Theory and Applications</u> (North Holland, Amsterdam).

McKibben, W. J. (1980) "Macroeconometric Models of the Australian Economy: A Comparative Analysis" <u>Research Discussion Paper, No. 8001</u> (Reserve Bank of Australia, Sydney, June).

Meeraus, A. (1983) "An Algebraic Approach to Modelling" <u>Journal of Economic Dynamics and Control</u>, Vol. 5, pp.81-108.

Miernyk, W. (1965) Elements of Input-Output Economics (Random House, New York).

Miernyk, W., et al. (1970) Simulating Regional Economic Development (Lexington Books, Lexington, MA).

Miernyk, W. H., (1973) "Regional and Interregional Input-Output Models: A Reappraisal" in Perlman, M., et al. (eds), Spatial, Regional and Population Economics (Gordon and Breach, New York), pp. 263-292.

Miller, R. (1966) "Interregional Feedbacks in Input-Output Models: Some Preliminary Results" Papers, Regional Science Association, Vol. 17, pp. 105-125.

Milne, W.J. (1981) "Migration in an Interregional Macroeconometric Model of the United States: Will Net Outmigration from the Northeast Continue?" International Regional Science Review, Vol. 6, No. 1, pp.71-84.

Milne, W.J., F.G. Adams and N.J. Glickman (1980) "A Top-Down Multiregion Model of the U.S. Economy" in Adams, F.G. and N.J. Glickman (eds) Modelling the Multiregional Economic System (Lexington Books, Lexington, MA).

Milne, W.J., N.J. Glickman and F.G. Adams (1980) "A Framework for Analysing Regional Growth and Decline: A Multiregional Econometric Model of the United States" Journal of Regional Science, Vol. 20, pp. 173-189.

MITI (Ministry of International Trade and Industry) (1970) "Interregional Input-Output Table for Japan" Trade and Industry in Japan, No. 108.

Monypenny, J.R. (1975) APMAA '74: Model, Algorithm, Testing and Application (APMAA Research #7, Department of Agricultural Economics and Business Management, University of New England, Armidale).

Moody, M. and F. Puffer (1970) "The Empirical Verification of the Urban Base Multiplier: Traditional and Adjustment Process Models" Land Economics, Vol. 46, pp. 91-98.

Moore, C. (1975) "A New Look at the Minimum Requirements Approach to Regional Economic Analysis" Economic Geography, Vol. 51, pp. 350-356.

Moses, L. (1955) "The Stability of Interregional Trading Patterns and Input-Output Analysis" American Economic Review, Vol. 45, pp. 803-832.

Moses, L. N. (1960) "A General Equilibrium Model of Production, Interregional Trade and Location of Industry" <u>Review of Economics and Statistics,</u> Vol. 45, pp. 373-397.

Mules, T.J. (1967) "Interindustry Analysis and the South Australian Wool Industry" Unpublished M. Econ. Thesis, University of Adelaide.

Muller, F. (1976) "An Integrated Regional Environmental-Economic Model" in P. Nijkamp (ed) <u>Environmental Economics</u> (Martinus Nijhoff, Boston, MA).

Muller, F. (1979) <u>Energy and Environment in Interregional Input-Output Models</u> (Martinus Nijhoff, Boston, MA).

Nijkamp, P. and J.H. Paelinck (1973) "An Interregional Model of Environmental Choice" <u>Papers, Regional Science Association</u>, Vol. 31, pp. 51-71.

Nijkamp, P. and P. Rietveld (1980) "Towards a Comparative Study of Multiregional Models" <u>Working Paper, WP-80-172</u>, International Institute for Applied Systems Analysis, Laxenburg, Austria, November.

Nijkamp, P. and P. Rietveld (1982a) "Structure Analysis of Spatial Systems" in B. Issaev, <u>et al</u> (eds) <u>Multiregional Economic Modelling: Practice and Prospect</u> (North-Holland, Amsterdam), pp. 35-48.

Nijkamp, P. and P. Rietveld (1982b) "Measurement of the Effectiveness of Regional Policies by Means of Multiregional Economic Models" in B. Issaev, <u>et al.</u> (eds) <u>Multiregional Economic Modelling: Practice and Prospect</u> (North-Holland, Amsterdam), pp. 65-81.

Nijkamp, P. and P. Rietveld (1982c) "Multiple Objectives in Multilevel Multiregional Planning Models" in M. Albegov, <u>et al.</u> (eds) <u>Regional Development Modelling: Theory and Practice</u> (North-Holland, Amsterdam), pp. 145-168.

Nijkamp, P., P. Rietveld and F. Snickars (1982) "Multiregional Economic Models: An Introduction to a Survey" in B. Issaev, <u>et al.</u> (eds) <u>Multiregional Economic Modelling, Practice and Prospect</u> (North Holland, Amsterdam), pp. 1-10.

Norman, A.L. and L.S. Lasdon (1983) "A Comparison of Methods for Solving and Optimizing A Large Non-Linear Econometric Model" <u>Journal of Economic Dynamics and Control</u>, Vol. 6, pp. 3-24.

O'Connell, M. (1981) "Regional Fertility Patterns in the United States: Convergence or Divergence?" <u>International Regional Science Review,</u> Vol. 6, No. 1, pp. 1-14.

Olsen, R.G. and G. W. Westley (1974) "Regional Differences in the Growth of Market Potentials, 1950-1970" Regional Science Perspectives, Vol. 4, pp. 99-111.

Olsen, R.G., G.W. Westely, et al (1977) Multiregion: A Simulation-Forecasting Model of B.E.A. Economic Area Population and Employment (Oak Ridge National Laboratory, ORNL/RUS-251, October).

Openshaw, S. (1976) "An Empirical Study of Some Spatial Interaction Models" Environment and Planning A, Vol. 8, pp. 23-42.

Orcutt, G. (1957) "A New Type of a Socio-Economic System" Review of Economics and Statistics, Vol. 58, pp. 773-797.

Orcutt, G., S. Caldwell, R. Wertheimer, et al (1976) Policy Exploration through Microanalytic Simulation (The Urban Institute, Washington, D.C.).

Orcutt, G., M. Greenberger, et al. (1976) Microanalysis of Socioeconomic Systems: A Simulation Study (Harper and Row, New York).

Pai, G. (1979) "Environmental Pollution Control Policy: An Assessment of Regional Economic Impacts" Unpublished PhD Dissertation, Massachusetts Insitute of Technology, June.

Park, S.H. (1970) "Least Squares Estimates of the Regional Employment Multiplier: An Appraisal" Journal of Regional Science, Vol. 10, pp. 365-374.

Parker, M.L. (1967) An Interindustry Study of the Western Australian Economy (Agricultural Economic Research Report No. 6, Perth: University of Western Australia).

Peterson, W., T. Barker and R. van der Ploeg (1983) "Software Support for Multisectoral Dynamic Models of National Economies" Journal of Economic Dynamics and Control, Vol. 6, pp. 109-130.

Pitfield, D.E. (1978) "Freight Distribution Model Predictions Compared: A Test of Hypotheses" Environment and Planning A, Vol. 10, No. 7, pp. 813-836.

Pitfield, D.E. (1979) "Freight Distribution Model Predictions Compared: Some Further Evidence" Environment and Planning A, Vol. 11, No. 2, pp. 223-226.

Plane, D.A. (1982) "An Information Theoretic Approach to the Estimation of Migration Flows" Journal of Regional Science, Vol. 22.

Plane, D.A. and P.A. Rogerson (1982) "Spatial Economic - Demographic Modelling of Interregional Migration with Limited Data" Paper presented at International Conference on Forecasting Regional Population Change and Its Economic Determinants and Consequences, Airlie, Virginia, May 26-29.

Polenske, K.R. (1970) "An Empirical Test of Interregional Input-Output Models: Estimation of 1963 Japanese Production" American Economic Reveiw, Vol. 60, No. 2, pp. 76-82.

Polenske, K.R. (1970a) "Empirical Implementation of a Multiregional Input-Output Gravity Model" in Carter, A.P. and Brody A. (eds) Contributions to Input-Output Analysis (North Holland, Amsterdam), pp. 143-163.

Polenske, K.R. (1972) "The Implementation of a Multiregional Input-Output Model of the U.S." in Brody, A. and A.P. Carter (eds) Input-Output Techniques (North Holland, Amsterdam), pp. 171-189.

Polenske, K.R. (1980) "A Methodology for Multiregional Economic Accounts, Policy Analyses and Forecasts" Paper presented at Conference on An Assessment of the State of the Art in Regional Modelling, Cambridge, Mass., April.

Polenske, K.R. (1980a) The U.S. Multiregional Input-Output Accounts and Model (Lexington Books, Lexington, MA).

Polenske, K.R., et al (1972) State Estimates of the Gross National Product, 1947, 1958, 1963 (Lexington Books, Lexington, MA).

Powell, A.A. and B.R. Parameter (1979) "The IMPACT Project as a Tool for Policy Analysis: A Brief Overview" The Australian Quarterly, Vol. 51, No. 1, pp. 62-74.

Pratt, R. (1968) "An Appraisal of the Minimum-Requirements Technique" Economic Geography, Vol. 44, pp. 117-125.

Preckel, P.V. (1985) "Alternative Algorithms for Computing Economic Equilibria" Mathematical Programming Study, Vol. 23, pp. 163-172.

Preston, R.S. (1972) "The Wharton Annual and Industry Forecasting Model" Studies in Quantitative Economics No. 7 (Dept. of Economics, University of Pennsylvania, Philadelphia, PA).

Redwood, A.L. (1982) "Forecasting Subnational Birth Rates" Paper presented at International Conference on Forecasting Regional Population Change and Its Economic Determinants and Consequences, Airlie Virginia, May 26-29.

Rees, P. and F. Willekens (1981) "Data Bases and Accounting Frameworks for IIASA's Comparative Migration and Settlement Study" Collaborative Paper CP-81-39, International Institute for Applied Systems Analysis, Laxenburg, Austria.

Rees, P.H. and A.G. Wilson (1973) "Accounts and Models for Spatial Demographic Analysis 1: Aggregate Population" Environment and Planning A, Vol. 5, No. 3, pp. 61-90.

Rees, P.H. and A.G. Wilson (1975) "Accounts and Models for Spatial Demographic Analysis 3: Rates and Life Tables" Environment and Planning A, Vol. 7, No. 2, pp. 199-232.

Richardson, H. (1972) Input-Output and Regional Economics (Weidenfeld and Nicolson, London).

Riefler, R.F. (1973) "Interregional Input-Output: A State of the Arts Survey" in Judge, G. and T. Takayama (eds) Studies in Economic Planning Over Space and Time (North Holland, Amsterdam).

Riefler, R.F. and C.M. Tiebout (1970) "Interregional Input-Output: An Empirical California - Washington Model" Journal of Regional Science, Vol. 10, No. 2, pp. 135-152.

Rietveld, P. (1981) "Causality Structures in Multiregional Economic Models" Working Paper, WP-81-50, International Institute for Applied Systems Analysis, Laxenburg, Austria, June.

Rietveld, P. (1982a) "A General Overview of Multiregional Economic Models" in B. Issaev, et al (eds) Multiregional Economic Modelling: Practice and Prospect (North Holland, Amsterdam), pp. 15-33.

Rietveld, P. (1982b) "A Survey of Multiregional Economic Models" in B. Issaev, et al (eds) Multiregional Economic Modelling: Practice and Prospect (North-Holland, Amsterdam), pp. 231-332.

Rodgers, J.M. (1972) State Estimates of Outputs, Employment and Payrolls, 1947, 1958, 1963 (Lexington Books, Lexington, MA).

Rodgers, J.M. (1973) State Estimates of Interregional Commodity Trade, 1963 (Lexington Books, Lexington, MA).

Roepke, H., D. Adams and R. Wiseman (1974) "A New Approach to the Identification of Industrial Complexes using Input-Output Data" Journal of Regional Science, Vol. 14, No. 1, pp. 15-29.

Rogers, A. (1966) "Matrix Analysis of Interregional Population Growth and Distribution" Papers, Regional Science Association, Vol. 18, pp. 177-186.

Rogers, A. (1973) "The Mathematics of Multiregional Demographic Growth" Environment and Planning A, Vol. 5, pp. 3-29.

Rogers, A. (1975) Introduction to Multiregional Mathematical Demography (Wiley, New York).

Rogers, A and P. Williams (1982) "A Framework for Multistate Demographic Modelling and Projection, with an Illustrative Application" Paper presented at International Conference on Forecasting Regional Population Change and Its Economic Determinants and Consequences, Airlie, Virginia, May 26-29.

Rose, A. (1974) "A Dynamic Interindustry Model for the Economic Analysis of Pollution Abatement" Environment and Planning A, Vol. 6, pp. 321-338.

Roseman, C.C. (1982) "Labour Force Migration, Non-Labour Force Migration and Non-Economic Reasons for Migration" Paper presented at International Conference on Forecasting Regional Population Change and Its Economic Determinants and Consequences, Airlie, Virginia, May 26-29.

Rowland, D.T. (1979) Internal Migration in Australia (Census Monograph Series, Australian Bureau of Statistics, Canberrra).

Saaty, T.L. (1981) Analytic Hierarchy Process: Planning, Priority Setting, Resource Allocation (McGraw Hill, New York).

Saaty, T.L. and M.W. Khouja (1976) "A Measure of World Influence" Journal of Peace Science, Vol. 2, No. 1, pp. 33-35.

Sakashita, N. (1982) "Recent Development of Multiregional Economic Models in Japan" in Issaev, B., et al (eds) Multiregional Economic Modelling: Practice and Prospect (North Holland, Amsterdam), pp. 189-202.

Saltzman, S. and H.S. Chi (1977) "An Exploratory Monthly Integrated Regional/National Econometric Model" Regional Science and Urban Economics, Vol. 7, Nos. 1/2, pp. 49-81.

Samuelson, P.A. (1966) "Non-Substitution Theorems" in Stigitz, I. (ed) The Collected Scientific Papers of P.A. Samuelson, Vol. I (MIT Press, Cambridge, MA), pp. 513-536.

Sasaki, K. (1963) "Military Expenditures and the Employment Multiplier in Hawaii" Review of Economics and Statistics, Vol. 45, pp. 298-304.

Scheppach, R.C. (1972) State Projections of the Gross National Product, 1970, 1980 (Lexington Books, Lexington, MA).

Schinnar, A.P. (1976) "A Multidimensional Accounting Model for Demographic and Economic Planning Interactions" *Environment and Planning A*, Vol. 8, No. 4, pp. 455-475.

Schinnar, A. P. (1977) "An Eco-Demographic Accounting-Type Multipler Analysis of Hungary" *Environment and Planning A*, Vol. 9, No. 4, pp. 373-384.

Schubert, U. (1982) "The Development of Multiregional Economic Models in Western Europe" in Issaev, B., et al (1982) *Multiregional Economic Modelling: Practice and Prospect* (North Holland, Amsterdam), pp. 99-109.

Schwartz, A. (1973) "Interpreting the Effect of Distance on Migration" *Journal of Political Economy*, Vol. 81, No. 5, pp. 1153-1169.

Shaw, R. P. (1975) *Migration Theory and Fact: A Review and Bibliograpy of Current Literature* (Regional Science Research Institue, Philadelphia, PA).

Shephard, R.W. (1953) *Cost and Production Functions* (Princeton University Press, Princeton, NJ).

Shryock, H., J.S. Siegel, et al (1976) *The Methods and Materials of Demography* (Condensed Edition by E. Stockwell) (Academic Press, New York).

Siegel, J. (1972) "Development and Accuracy of Projections of Population and Households in the United States" *Demography*, Vol. 9, pp. 51-68.

Smith, C. , 1983, Integration of Multiregional Models for Policy Analysis, Unpublished Ph.D. Dissertation, Cornell University, January.

Smith, C., 1986a, "An Empirically Implementable Multiregional Model for Australia" *Regional Science and Urban Economics*, Vol. 16, pp. 181-195.

Smith, C., 1986b "Is There an Alternative to ORANI for State-Based Economic-Demographic Projections" Paper presented at Economist's Conference, Melbourne, August.

Smith, C., 1986c, "Development of a Multiregional Econometric Model for Australian States" Paper presented to the Input-Output Workshop at the Eleventh Annual Meeting of the Australian and New Zealand Section of the Regional Science Association, Sydney, December.

Smith, T.E. (1978) "A Cost-Efficiency Principle of Spatial Interaction Behaviour" *Regional Science and Urban Economics*, Vol. 8, No. 4, pp. 313-337.

Smith, T.E. (1981) "A Cost Efficiency Approach to the Analysis of Congested Spatial Interaction Behaviour" *Working Papers in Regional Science and Transportation*, No. 55, October.

Snickars, F. (1978) "Estimation of Interregional Input-Output Tables by Efficient Information Adding" in C. Bartels and R. Kettellapper (eds) *Exploratory and Explanatory Analysis of Spatial Data* (Martinus Nijhoff, Lieden).

Snickars, F. (1981) "Interregional and International Linkages in Multiregional Economic Models" Paper presented at International Conference on Structural Economic Analysis and Planning in Time and Space, Umea, Sweden, June.

Snickars, F. (1982a) "The Regional Development Consequences of a Shutdown of Nuclear Power Plants in Sweden" in M. Albegov, et al (eds) *Regional Development Modelling: Theory and Practice* (North Holland, Amsterdam), pp. 361-375.

Snickars, F. (1982b) "Interregional Linkages in Multiregional Economic Models" in B. Issaev, et al (eds) *Multiregional Economic Modelling: Practice and Prospect* (North Holland, Amsterdam), pp. 49-64.

Snickars, F. and A. Granholm (1981) "A Multiregional Planning and Forecasting Model with Special Regard to the Public Sector" *Regional Science and Urban Economics*, Vol. 11, No. 3, pp. 377-404.

Snodgrass, M.M. and C.E. French (1958) *Linear Programming Approach to Interregional Competition in Dairying*, Purdue University, Agricultural Experiment Station, LaFayette, Indiana, May.

Sommerfeld, J.T., et al. (1977) "Identification and Analysis of Potential Chemical Manufacturing Complexes" *Journal of Regional Science*, Vol. 17, no. 3, pp. 421-430.

Spilerman, S. (1972) "The Analysis of Mobility Processes by the Introduction of Independent Variables into a Markov Chain" *American Sociological Review*, Vol. 37, pp. 277-294.

Steed, X. (1970) "Changing Linkages and Internal Multipliers of an Industrial Complex" *Canadian Geographer*, Vol. 14, pp. 229-242.

Steenge, A.E. (1977) "Economic-Ecologic Analysis: A Note on Isard's Approach" *Journal of Regional Science*, Vol. 17, no. 1, pp. 97-105.

Stevens, B.H. (1958) "An Interregional Linear Programming Model" *Journal of Regional Science*, Vol. 1, No. 1, pp. 60-98.

Stillwell, J.C.H. (1978) "Interzonal Migration: Some Historical Tests of Spatial Interaction Models" *Environment and Planning A*, Vol. 10, No. 10, pp. 1187-1200.

Suzuki, N., F. Kimura and Y. Yasuyuki (1978) <u>Regional Dispersion Policies and Their Effects on Industries--Calculations Based on an Interregional Input-Output Model</u> (Mitsubishi Research Institute, Tokyo).

Taylor, J.C. (1979) "Some Aspects of RBA 76 and RBF 1" <u>Research Discussion Paper No. 7904</u> (Reserve Bank of Australia, Sydney, September.

Tiebout, C.M. (1957) "Regional and Interregional Input-Output Models: An Appraisal" <u>Southern Economic Journal, Vol. 24, pp. 140-147.</u>

Tiebout, C.M. (1962) <u>The Community Economic Base Study</u> (Committee for Economic Development, New York).

Todaro, M.P. (1976) "Urban Job Expansion, Induced Migration and Rising Unemployment" <u>Journal of Development Economics</u>, Vol. 3, pp. 211-225.

Treyz, G.I. (1980) "Design of a Multiregional Policy Analysis Model" <u>Journal of Regional Science,</u> Vol. 20, No. 2, pp. 191-206.

Treyz, G.I. and B.H. Stevens (1980) "Locational Analysis for Multiregional Modelling" in Adams, F.G. and N.J. Glickman (eds) <u>Modelling the Multiregional Economic System</u> (Lexington Books, Lexington, MA), pp. 75-88.

Treyz, G.I., A.F. Friedlaender and B.H. Stevens (1980) "The Employment Sector of a Regional Policy Simulation Model" <u>Review of Economics and Statistics,</u> Vol. 62, pp. 63-73.

Varian, H.R. (1978) <u>Microeconomic Analysis</u> (Norton, New York).

Veen, van der, A. and G.H.M. Evers (1982) "Labour Force Participation, Migrating, Commuting and the Regional Labour Market: A Simultaneous Approach" Paper presented at International Conference on Forecasting Population Change and Its Economic Determinants and Consequences, Airlie, Virginia, May 26-29.

Victor, P. (1972) <u>Pollution, Economy and Environment</u> (Allen and Unwin, London).

Wachter, M. (1975) "A Time-Series Fertility Equation: The Potential for a Baby Boom in the 1980's" <u>International Economic Review,</u> Vol. 16, No. 3, pp. 609-624.

Wadycki, W. (1979) "Alternative Opportunities and the United States Interstate Migration: an Improved Econometric Specification" <u>Annals of Regional Science,</u> Vol.13, No. 3, pp. 35-41.

Walker, N. and J. Dillion 91976) "Development of an Aggregative Programming Model of Australian Agriculture" <u>Journal of Agricultrural Economics</u>, Vol. 27, No. 2, pp. 243-248.

Weber, A. (1929) <u>Theory of the Location of Industry</u> (translated by C. Friedrich) (University of Chicago Press, Chicago, Ill).

Weiss, S. and E. Gooding (1968) "Estimation of Differential Employment Multipliers in a Small Regional Economy" <u>Land Economics</u>, Vol. 44, pp. 235-244.

West, G.R. J.T. Wilkinson and R.C. Jensen (1979) <u>Generation of Regional Input-Output Tables for the State and Regions of South Australia</u> (Report to the Treasury Department, the Department of Urban and Regional Affairs and the Department of Trade and Industry; Dept. of Economics, University of Queensland, St. Lucia).

West, G.R. J.T. Wilkinson and R.C. Jensen (1980) <u>Generation of Regional Input-Output Tables for the Northern Territory</u> (Report to the Northern Territory Dept. of the Chief Minister; Dept. of Economics, University of Queensland, St. Lucia).

Wicks, J.A. and J.L. Dillion (1978) "APMAA Estimates of Supply Elasticities" <u>Review of Marketing and Agricultural Economics</u>, Vol. 46, No. 1, pp. 48-57.

Wharton Econometric Forecasting Associates (1982) <u>The Wharton Annual Model Extension to the Year 2001</u>, W.E.F.A., Philadelphia, February.

Wilen, J.E. (1973) "A Model of Economic System - Ecosystem Interaction" <u>Environment and Planning A</u>, Vol. 5, pp. 409-420.

Willekens, F. (1980) "A Simulation Procedure for Multiregional Population Analysis and Forecasting" Paper presented at First World Regional Science Congress, Cambridge, Mass., June.

Williams, I. (1976) "A Comparison of Some Calibration Techniques for Doubly Constrained Models with an Exponential Cost Function" <u>Transportation Research</u>, Vol. 10, pp. 91-104.

Wilson, A.G. (1970) <u>Entropy in Urban and Regional Modelling</u> (Pion, London).

Wilson, A.G. and P.H. Rees (1974) "Accounts and Models for Spatial Demographic Analysis 2: Age-Sex Disaggregated Populations" <u>Environment and Planning A</u>, Vol. 6, pp. 101-116.

Wilson, A.G., P.M. Rees, and C.M. Leigh (eds) (1977) <u>Models of Cities and Regions</u> (Wiley, New York).

Woodstock, J.T. (ed) (1980) <u>Mining and Metallurgical Practices in Australasia</u> (The Australasian Institute of Mining and Metallurgy, Monograph Series, No. 10).